THE CODE

THE CODE

AN AUTHORIZED HISTORY
OF THE ASME
BOILER AND PRESSURE VESSEL CODE

Wilbur Cross
Illustrations by Judy Pensky Alter

The American Society of Mechanical Engineers

New York

Reprinted 2020.

Library of Congress Cataloging-in-Publication Data

Cross, Wilbur.
   The code : an authorized history of the ASME boiler and pressure vessel code / Wilbur Cross,
      p. m.
      ISBN 0-7918-2024-6
      1. Steam-boilers — Standards — United States — History. 2. Pressure vessels — Standards — United States — History. I. American Society of Mechanical Engineers. II. Title.
TJ288.C76 1989
621.1'83'0218-dc20   89-18037
                                                          CIP

CONTENTS

v

PREFACE
Viewpoint

Over a period of many months, my editorial colleagues and I have had welcome opportunities for discussions with engineers of all ages and in many locations about the history of the profession — where it has been, where it is now, and where it is going. Some of the most provocative of these lively dialogs have been with young graduate engineers who were not only enthusiastic about the work they were engaged in, but who were still close enough to their scholastic origins to recall quite clearly why it was that they selected engineering for their career and lifetime commitment.

"I have always been fascinated with dams and the control of one of nature's most powerful and productive forces," said a young man who was excited about his prospects of being able to specialize in this field.

"Originally, when I was still in high school," admitted a female graduate, "I intended to become an architect for I was always motivated by the fact that this career would permit me to satisfy my love of beauty on the one hand while exercising my practical bent on the other. But along the way I became even more fascinated by the concept of bridges, ranging all the way from the antique covered variety and graceful stream crossings to the dramatic sweep of suspensions across bays and wide rivers."

"You may think I am one of those people who likes to hide his light behind a barrel," explained yet another young graduate, "when I tell you that my field of endeavor relates to something few people ever see: foundations. I intend to become a specialist in foundations for tall structures and also to make a study of earthquakes to determine how we can design structures and power plants that will be indestructible even when threatened by the greatest tremors."

It was encouraging and stimulating to talk to these young people and learn why they had been drawn to engineering in general and their respective fields in particular. Although a non-engineer, I could easily understand what had motivated these graduates to begin with and what excited them most about their promising careers. As a writer, I had always found it easy to equate emotional and poetic values with tall buildings, gracefully arching spans, or power plants with the capability of changing both the geographical and economic faces of vast regions and even whole nations.

These were the engineering ultimates, the visual evidence that engineering was one of the great disciplines when it came to the act of making mankind's dreams come true and turning wishes into realities. They were the subjects captured on canvas by painters, extolled in sonnets by poets, dramatized in novels by authors, and in not a few instances popularized in song by musicians.

And then I met a different breed of young engineer, one who gave me pause to think and reflect. "I am a mechanical engineer," he reflected, "one of those types who has a desire — almost an obsession — to see the products of the creative engineering mind run a little bit more efficiently, a little more durably, and a great deal safer than has been possible in the past, whether distant or recent.

"If I were asked what engineering achievements in the last century stand out as being the most remarkable, that is why I would reply quite differently from most of my peers. I would give the 'Oscar' to the Boiler and Pressure Vessel Code. And in so doing, I would not be singling out a lone innovator, a genius standing on stage. I would be recognizing an army of engineers who had the foresight to realize a need, the guts to take on a very frustrating challenge, and the durability to shed the comments of critics and stick with the jobs that had to be done."

Most importantly, he added, the labors and achievements of Code committees over the years represented the finest example of teamwork he had ever studied. He could think of no other program that reflected such a common and wholehearted desire to ensure that product quality was uniformly high and that the user was well served. Even for one who had devoted several years to the study of mechanical engineering, he commented, it was impressive to understand what properly oriented codes and standards could accomplish in the matter of faster installation, operational efficiency, durability, and safety.

If this published history does no more than make the profession and the public aware of how the Code has enhanced the quality of life, it will have accomplished an important mission.

The Author

ACKNOWLEDGMENTS

In recognition by its very nature, the ASME Boiler and Pressure Vessel Code has always relied on committees and other groups, rather than on single individuals, to create, organize, and distribute codes and standards and administer the many functions relating to them. Any acknowledgment of contributions has to recognize this fact and pay tribute to the unnamed thousands of people who have developed the Code during its 75-year history, and, going back earlier, to those numerous pioneers who sensed a need in the first place and were instrumental in setting procedures in motion that would eventually lead to success.

To that end, this book is dedicated to those people who have made such a history possible, from the participants who undertook the initial spade work in the years 1911 through 1914, to the committee responsible for the publication of the first Code in 1914 and all those that succeeded it in the years that followed to the present day. It is to their common qualities of foresight, determination, dedication, and objectivity that the Code was able not only to become a reality, but to take its rightful place as one of the great triumphs in the field of mechanical engineering.

One of the constant challenges in writing the Code was — and is to this day — the need to anticipate technological innovations, revolutions, and trends far enough in advance so that the rules and specifications would be applicable at the time the changes were taking place and not after a considerable time lag for adjustments. This book serves as recognition of the engineers and other professionals whose talents and sound judgments have made the Code what it is and established those qualities for which it stands.

While it is all but impossible to list individual pioneers and contributors to the Code without leaving out multitudes who should also be so recognized, it is one of the rewards of authorship that a writer traditionally is given license to name names and thank those who have made his assignment more pleasant, his work more effective, and his voice more articulate. There is little doubt that Frank S. G. Williams should be commended for his long and untiring efforts to compile historical material relating to the Code, starting many years before the author had the pleasure of working with him. And from start to finish, Frank was a guiding light and cheerful collaborator, as well as a strict disciplinarian when it came to checking facts and seeking authentication.

The author is almost equally indebted to Walter Harding who devoted many days to reviewing proposed chapters for the book, correcting errors and misleading statements, and making recommendations for additions and revisions. Leonard P. Zick was also helpful in reviewing chapter drafts, as well as submitting tear sheets and reports on various topics pertinent to the history. Others who supplied information and materials that were particularly welcome and appreciated were Howard Dobel and his colleagues, who provided a wealth of historical data from the archives of the Hartford Steam Boiler & Inspection Company; Patrick Moore of *Materials Evaluation* with a fine history of nondestructive testing; A. F. Manz, who suggested reference sources for a history of welding; James Andrew, who not only provided information and sound advice but pointed out several serious omissions in an early draft of the book; James Clarke and Paul Brister, who collaborated with Frank Williams on the initial historical research; Euan Somerscales, who has become an authority on ASME's historical landmarks; John Clough and his associates at Babcock & Wilcox for impressive copies of *Steam* and other historical publications; and John Harvey for documents relating to the diversity of Code formulas.

The author concludes with special thanks to all of those ASME staff members in Codes and Standards who have made his assignment not only easier but more enjoyable, with particular recognition to Bill Woollacott and Mel Green for encouraging him to write objectively as well as authoritatively in a manner befitting the subject itself.

CHAPTER 1

CODES IN ACTION
Recent Highlights and Milestones

In the spring of 1925, a group of mechanical engineers gathered in Weymouth, Massachusetts, a quiet residential community just southeast of Boston, to watch the President of the Boston Edison Company pull a small switch at what had been designated the Edgar Steam-Electric Station. It was not a particularly unusual scene in the flourishing Northeast, where electric stations were springing up all over and where steam power had become a kind of symbol of "The New Industrial Revolution" that was taking place. Massachusetts was at the hub of the power breakthrough, with its burgeoning shoe factories, woolen mills, and paper plants. In this area, the Boston Edison Company was at the forefront of the power boom.

In the minds of the engineers, however, the high-pressure topping turbine and boiler at Edgar represented something that could be more innovative than anything which had preceded them. Their expectations and aspirations proved to be right on target. From the moment it went on stream and during the rest of the Roaring Twenties, the Edgar Steam Electric Station set new records for economy. It achieved the unheard-of goal of producing electricity at the rate of 1 kilowatt hour per 1 pound of coal. Most surprising, this occurred at a time when conventional power plants were readily competing with each other by consuming five to ten pounds per kilowatt hour.

Boston Edison achieved this feat by operating the boiler and turbine unit at 1,200 pounds of steam pressure and exhausting into a 350-pound steam header — in its day the only operation of its kind in the world. Another "first" was the X-raying of the unit's steel piping and turbine casings to ensure a flawless subsurface, a procedure that became standard practice, though not until much later. Most surprising of all, perhaps, was the fact that the station stood as a monument to the efficiency of its designers and engineers until the late 1970s when it was dismantled and shipped to a power company in South America.

On May 26, 1958, another switch was tripped during another historic moment in the development of power, this time by President Dwight David Eisenhower in the White House in Washington, D.C. His action sent an impulse across the miles to Shipping port, a town on the Ohio River northwest of Pittsburgh, where a second switch was activated. The occasion was the dedication of the Shipping port Atomic Power Station, the first commercial central electric-generating station in North America to utilize nuclear energy. Designed and built for the Department of Energy and the Duquesne Light Company, this historic station at Shipping port was a key element in the Federal administration's "Atoms for Peace" program, designed to transform what had been considered a monster of war into a white knight for the future of mankind.

Designed by the Westinghouse Electric Corporation in cooperation with the Division of Naval Reactors of the Atomic Energy Commission, the Shipping port facility was relatively small by later standards, generating only 60,000 kilowatts of electricity.

In the early 1920s in Sonoma County, California, north of San Francisco, two young men with a vision were fascinated by a sight they saw for the first time — billows of steam that were escaping from fumaroles in a natural geothermal field. If the steam could be harnessed, they reasoned, it could supply never-failing amounts of heat and power to communities in the area, the number being limited only by the extent to which a network of insulated piping could be constructed.

In 1922, a small company was formed, plans were laid, and equipment was brought into the geothermal field to capture the natural energy that was so visible all around. But the steam and hot water erupting from the earth were far too corrosive for the piping and other materials available at the time. Anti-corrosion technology had not yet advanced to the point where the problems could be overcome, and the venture had to be abandoned. Nevertheless, the idea of harnessing this substantial and dependable force of nature continued to intrigue the imagination, as it had since the discovery of the Sonoma field in 1847.

In the late 1950s, the scene was again one of activity and by 1960, Geysers Unit #1 was in actual operation, as the first commercial geothermal electric-generating station in North America. Basically the unit was nothing more complex than a rebuilt kilowatt-size General Electric generator installed by Pacific Gas and Electric Company engineers in such a way that it made use of the natural geothermal energy instead of commercially produced steam. From this initial turbogenerator grew the largest geothermal development in the world, pioneering the design of systems to provide corrosion-resistant cooling, remove noncondensible gas, and incorporate reliable environmental controls.

The Geysers became known as the largest vapor-dominated geothermal system, surpassing those of Lardello, Italy, which had been the world's first geothermal installation for power generation. Providing fairly clean, slightly superheated dry steam, the California field proved to be ideal for electricity generation in that the steam rose by natural means and could be applied directly to the turbine. Once in full operation, Geysers Unit #1 regularly began providing 665,000 net cumulative kilowatts from 13 generators tapped into more than 200 steaming wells.

No one would ever question the fact that projects like these in the field of power and energy come into being only through countless years of work, combinations of the highest technical skills, and impressive accumulations of engineering experience extending back through many decades. Yet what the general public is not aware of, and what escapes all but a small segment of

4

even the most educated, is the fact that innovative projects like these demand a comprehensive and very intricate cumulation of codes, standards, rules, and specifications that have literally taken generations to evolve.

Such codes and standards may not necessarily be applied during the pioneering stages of development, when ingenuity and invention are playing the key roles, but in due course of time they become all-important factors. And they, too, may often serve as catalysts in the improvement of the metals, materials, components, and products whose integrity and utility they are protecting. The very history of codes and standards — and most specifically of the ASME Boiler and Pressure Vessel Code — demonstrates this kind of progression upward and onward. In essence, the Code has come to assure designers, manufacturers, owners, users, insurers, and the general public that a unit of engineering equipment will operate safely and reliably if it is designed in accordance with the rules and specifications and able to meet the regular examinations of qualified inspectors. Over and beyond that, the Code has served historically not only as a benchmark but as a springboard to higher quality and better performance.

A significant effort by the ASME in the field of standardization came in 1902 when the Council appointed a Special Committee for Standard Proportions for Machine Screws. The results of the investigation that followed were published five years later. Shortly after that the Society began its study of codes, climaxed by the creation of the Boiler Code in 1914.

The ASME Council on Codes and Standards was formed and evolved into a fairly complex structure which, by the mid-1980s, had five Supervisory Boards and five Advisory Boards reporting to it, each functioning within specific subject areas. The Board on International Standards is one such body, the responsibilities of which include recommending policies affecting international standardization, reviewing proposed activities in the area of international standardization to determine whether they fall within the scope of the Council, and proposing the allocation of funds in connection with international conferences on standards that are scheduled in the United States.

In all activities, an underlying responsibility of the Council has been to see to it that ASME procedures and those of the American National Standards Institute are carefully followed, in spirit as well as in structure.

Since its founding in 1880, the American Society of Mechanical Engineers has emphasized as its stated purpose the stimulation of technical exchange, accomplished through the conventional media of conferences and exhibits and through the printed page in the form of pamphlets, periodicals, and books, the subjects of which evolve in many instances from research and the writing of professional papers by members and others qualified to communicate in their fields of technical and industrial endeavor.

5

A Society activity of prime importance has been the development and promulgation of sound engineering standards and safety codes, a notable example of which is the ASME Boiler and Pressure Vessel Code, which is referenced in safety regulations of major cities, states, provinces, and other such governments in the United States and Canada. In addition, a number of top government agencies include the Code, or portions of it, in their respective regulations.

In the administrative structure of the ASME, the Committee answers to the Board on Pressure Technology Codes and Standards for administrative and technical matters regarding boilers and pressure vessels and to the Board on Nuclear Codes and Standards for technical matters regarding nuclear codes and standards.

The ASME Boiler and Pressure Vessel Code was conceived and perfected as a technical document to communicate the minimum requirements for the design, fabrication, installation, and inspection of boilers and pressure vessels and related components. Its preparation requires the services of people knowledgeable about the processes and equipment involved in the components covered by the rules and specifications. The groups assigned specifically to the writing of the Code are composed of competent designers, fabricators, users, and others with general or specialized interests. The general interest group, for example, might include representatives in the fields of consulting, education, research, or government agencies. Those with specialized knowledge and capabilities might include metallurgists, atomic-energy scientists, chemists, or geophysicists.

For a code to receive wide respect and acceptance, it must embody the latest and best knowledge in those disciplines within its scope. This has been the prime objective of the Boiler and Pressure Vessel Code from the beginning and has been reflected increasingly in its recognition as a *voluntary* standard by manufacturers and users throughout the world. What this means is that ASME, through its internal membership, has voluntarily furnished the technical knowledge of design, materials, and fabrication necessary to define standards of safety. Once that has been done, the regulatory authorities in the cities, states, and other regional units that adopt the Code for local reference have made certain that those within their jurisdiction meet the requirements of the specifications and rules established. A Code has mandatory status only when it is authorized and enforced by some political entity.

In a recent public statement, the Society reiterated its position that its "overriding objective is to play an even more active and effective role in the generation and the continuous improvement of codes and standards under the Society's jurisdiction or in cooperation with other agencies, both domestic and international, and to increase our efforts toward the implementation of such codes. The purposes of codes, standards, and related accreditation programs are to enhance public health, safety, and welfare."

The function of the Code Committee is to seek out, review, and develop the criteria needed to address the scope and purpose of each rule or specification, as well as to interpret elements of the Code whenever questions arise as to their meaning and content. In the United States, there is no one specific governmental agency or unit that establishes standards of this nature. Rather, each governmental body writes into its own laws those codes or standards that best provide a means of satisfying regulatory requirements. The overall coordination of the American system is traditionally undertaken by the American National Standards Institute, which oversees the development and approval of national standards. ANSI, in turn, is the United States member of such groups as the International Organization for Standardization (ISO), the International Electrotechnical Commission (IEC), and the Pacific Area Standards Congress (PASC). ASME is also a member of ANSI and is accredited by ANSI to develop American National Standards. ANSI based its accredited organization procedures on ASME procedures after the Hydrolevel case. The Boiler and Pressure Vessel Committee procedures were the first to be accepted under the organization method adopted in 1972. Acceptance resulted in a study of ANSI's structures and purposes and two new methods of developing standards — the Accredited Organization Method and the Accredited Procedures Method. Furthermore, the Canvass Method was retained and the Committee Method was eliminated. ANSI approval of a standard verifies that the principles of due process and objectivity have been followed in the approval procedure and a consensus of those who are directly and materially affected by the standard has been achieved.

From its inception, the Boiler and Pressure Vessel Code Committee has used several practical avenues for public participation and comment. One of these avenues has been the announcement of all proposed revisions and new sections in *Mechanical Engineering,* ASME's magazine, and ANSI's *Reporter.* Another is the scheduling of public hearings to hold open discussions of the proposed changes or additions. Such hearings are advertised in advance in national newspapers, technical journals, and regional and local newspapers as well. All such meetings are open to members of the public, whether they belong to the ASME, other professional associations, or none at all.

Once agreement is reached on additions or revisions, proposed Code changes (or abstracts if the text is extensive) are published in *Mechanical Engineering* (circulation approximately 120,000), and their availability is announced in ANSI's *Reporter* (sent to 10,000 readers), to allow additional time for public review and comment. All responses, in whatever form, are considered by the Committee before, or during, the next meeting. Once agreement has been reached, the Committee submits its proposed revisions to the appropriate ASME supervisory board, as well as to the American National Standards Institute. When both of these bodies have given approval, the revisions are incorporated into the Code and become part of an American National Standard.

In those rare situations where comments cannot be resolved by the Committee, the traditional alternative has been to establish a system of appeals and hearings. Viewed in sequence, the subjects are reviewed first by the appropriate subcommittee, then by the main committee, the supervisory board, and finally the Board on Hearings and Appeals.

In all instances of forming or modifying codes and standards, from start to finish, the procedure follows the course of due process of law. This is defined as "the regular administration of the law, according to which no citizen may be denied his or her legal rights and all laws must conform to fundamental, accepted legal principles." In the work of the Code committees, this has become a kind of administrative slogan, simply referred to as "due process."

Besides setting standards and constantly modifying and updating the rules therein, the Code Committee also administers a system of accreditation for manufacturers of equipment. This system has been functioning since the publication of the first Boiler Code in 1915. The objective of accreditation, in this instance, is to provide a means for distinguishing equipment that meets Code requirements from that which may not meet such requirements. Accreditation is to provide for the identification of manufacturers whose quality assurance systems have been reviewed to establish that they have the capability to construct equipment in compliance with Code rules. Manufacturers so accredited are authorized to apply the ASME Code Symbol Stamp as their certification that individual products have been constructed to Code rules. This symbol distinguishes Code-built equipment from products not so marked.

Although derived from the American legal system, the ASME procedure is applicable on an international basis. The policy was strengthened when, in 1972, a consent decree was uniformly approved by the Society, the United States Government, and the National Board of Boiler and Pressure Vessel Manufacturers, which mandated "equal treatment of American and foreign manufacturers as to the use of our certification stamps and plant inspections." It further required that "all fees be computed on the same basis as in the case of domestic manufacturers and inspectors," emphasizing that the procedures used to accredit foreign manufacturers had to be similar to those applied to domestic manufacturers, subject to modifications necessitated by distance, language, and units of measurement. Once a standard was approved, the

8

question had to be asked, "Is that standard acceptable in a country that has a different set of operations and controls?"

"What is the Boiler and Pressure Vessel Code?"

For anyone who asks this question, a small ASME booklet has a brief answer: "The Boiler and Pressure Vessel Code establishes rules of safety governing the design, fabrication, and inspection of boilers and pressure vessels, and nuclear power plant components during construction. The objective of the rules is to assure reasonably certain protection of life and property and to provide a margin for deterioration in service. Advancements in design and material and the evidence of experience are constantly being added by Addenda."

The implications and substance of the Code were summed up more than 35 years ago in a one-sentence statement by The American National Standards Institute, then known as the American Standards Association: *"Probably no other standard in America has done more for the national safety."*

In January 1984, the ASME published a 16-page brochure entitled "The Why and How of Codes and Standards," with an introduction by William E. Cooper, then Senior Vice President, Codes and Standards. Although this modest-sized, illustrated publication was meant to be no more than a quick overview of the subject, the brochure presented a telling and provocative interpretation of the origins and concepts of the original Boiler Code, even before it became an actuality. Speaking of that era shortly after the creation of ASME in the 1880s, the text explored "one of the more innovative ideas of the 19th century," producing steam and then converting it into energy to power machinery.

"For want of reliably tested materials, secure fittings and proper valves, boilers of every description, on land and at sea, were exploding with terrifying frequency. They would continue to do so into the 20th century. Engineers could take pride in the growing superiority of American technology but they could not ignore the price of 50,000 dead and two million injured by accidents annually."

The mechanical engineers who tackled the problems of steam power began by seeking reliable methods for testing boilers, but were hindered by the fact that universally accepted construction standards lay many years in the future. "It took time and circumstance, as well as persuasion and common sense, to establish a system of voluntary standardization. To cut down chaos and high cost, the interchangeability of parts had to assert itself. . . . Mechanical engineers confronted a whole range of design and manufacturing tasks, including pipe and pipe threads, cast-iron water and gas pipe, boiler tubes, gas burners, hose couplings, bolt heads and nuts, pumping engines, marine and factory engines, gas and oil engines, and many more.

9

"Because technology never stands still," concluded the brochure, "ASME realized that it would be necessary to continually update and revise most of its codes and standards to keep pace with new materials, new designs, and new applications. . . .

"ASME never assumes that a Code is the last word — only the *latest* word."

How did this all come about?

The story goes back to man's earliest concepts of hot water and steam as elements that could be harnessed by humans to improve living standards, accomplish great works, and change the course of history.

CHAPTER 2

GENERATIONS PAST

Early History of Steam Power

As far back as the Middle Ages, engineers were consumed with the challenge of developing non-human forms of power that would ease the burdens of workers while at the same time making available more comforts and conveniences for a greater number of people. The major sources of such power were animals, wind, water, and other natural forces, harnessed to drive an increasing variety of devices and inventions, many of them remarkably ingenious and efficient. Modern machinery derives in part from the primitive quern, which was used for grinding grain, and applied the force of two millstones that rotated upon each other. Gears and cogwheels were utilized to increase the power of machines operated by non-human forces. And more sophisticated variations of the ancient windmill demonstrated engineering innovations like tilted sails to catch the wind more effectively, and multiple axles and gears that would utilize more power while driving pumps and other machinery.

Although many engineers and inventors had dreams of harnessing steam in past centuries, it remained an elusive force. Two men of Hellenistic times, Ctesibius and Hero, were renowned for their investigations into the properties of steam produced by boiling water and wrote descriptions of pistons, pumps, and engines that could one day function on behalf of mankind. By the early 17th century, one Giovanni Battistadella Porta, who had studied the writings of Hero, discovered that when steam condensed in a closed vessel it created a vacuum that could be used to draw up water, a principle that would be experimented with in a steam engine a hundred years later.

The practical application of steam had to wait, however, until an Englishman, Thomas Savery, invented and patented a steam pump for draining water out of coal mines. His invention had no moving parts except cocks and check valves, which were turned by hand to let steam into a closed vessel. When chilled from the outside by cold water, the vessel condensed the steam, creating a vacuum strong enough to draw water upward through a suction pipe. The water was then flushed away by the further introduction of steam. In spite of its inefficiency, the pump could complete each cycle of operation in about 12 minutes.

Savery's engine marked the historic advent of the steam engine, providing a springboard for Thomas Newcomen a decade later when he introduced his own engine for pumping water. Yet, it too was slow and inefficient and consumed huge amounts of coal by comparison with the workload it accomplished. Critics, pointing out that the wastage of fuel was more than 90 percent, liked to declare that "it takes an iron mine to build a Newcomen engine and a coal mine to keep it going."

The Newcomen machine also provided inspiration for James Watt, a Scottish engineer who had noted the waste of heat between strokes in a

Newcomen engine he was repairing, and was convinced that he could make improvements. In his design, he continued to utilize steam and a partial vacuum, but he provided a separate chamber for condensing the steam and, with other modifications, cut fuel consumption to about 25 percent of what previous models had required.

By the end of the 18th century, the use of steam had become so common that a standard had to be devised to judge the capacities of engines. It was during this period that "horsepower" came into common usage. It was literally derived from the comparison of the work capacity of a steam engine with what could have been accomplished by a horse instead of steam. Watt had calculated that a brewery horse could produce 33,000 foot pounds per minute — that is, the mechanical force to lift 33,000 pounds 1 foot in 1 minute. By 1809, this was accepted as *one horsepower,* the measurement that is still in use today.

Of particular importance to the history of boilers was the work of Richard Trevithick, who grew up in the Cornish mining district of England and from an early age on had become acquainted with steam engines, pumps, and other machinery. He became a self-taught engineer and experimented with steam engines, realizing as he did so that one of the major problems in steam systems was the construction of the boiler itself. In 1800, he designed an engine that operated at a pressure of 65 pounds per square inch. The boiler and engine were mounted together, representing a milestone in the history of boilers. The design for the single pipe flue that was then used was replaced by more and more tubes, essentially increasing the heating surface and adding to the efficiency of the boiler. However, fire-tube boilers, which were limited in the pressures they could withstand as well as their capacity, were not destined to fulfill the demands in the middle and late 19th century for larger units and higher pressures.

An American engineer, John Stevens, had constructed a boiler with multiple tubes as early as 1825. He was also one of the first to design units in which tubes carried, not the gases from the fire, but the water that was to be turned into steam. His first "water-tube" boiler consisted of a group of small tubes closed at one end and connected at the other to a central reservoir. One such boiler was used for a short time to supply steam for the engine of a small Hudson River steamboat. Although the vessel's service was short-lived because of mechanical problems, the patents that the inventor persuaded Congress to register on his behalf were said to have formed the basis of laws that to this day protect innovative designs for water-tube boilers.

It was not until three decades later that Stephen Wilcox perfected what was later considered a "real breakthrough" in water-tube boilers. He designed inclined water tubes that connected water spaces at the front and rear, with a steam chamber above. This permitted much better circulation of the water, provided more heating surface, and reduced the risk of explosion that was one of the banes of the early water-tube design.

While boilers were undergoing significant changes, steam engines were also being improved steadily to meet the growing demands of industrialization on both sides of the Atlantic. New designs for stationary reciprocating steam engines saw increases in size, speed, and efficiency, making them much more economical and competitive. In 1849, George Henry Corliss, of Providence, Rhode Island, invented a unique quick-acting valve to replace the slide valve then in general use on stationary engines. Consequently, the Corliss engine achieved great popularity, not only in the United States but in Europe, where manufacturing industries were springing up in all of the major cities. A giant-sized Corliss engine was one of the most notable exhibits at the 1876 Centennial Exhibition in Philadelphia, where it dwarfed most of the other splendors in Machinery Hall.

The Exhibition was the first world's fair at which the practicality of generating electricity by steam power was demonstrated to the public. From this time on, the use of steam power in mills and factories was to become commonplace. The next three decades saw the gradual shifting from the early "shell" boiler, which was little more than a kettle filled with water and heated at the bottom, to the "fire-tube" type with heat and hot gases inside the tubes, and then to the "water-tube" design, which eventually was considered to be reasonably safe and efficient. Along the way, drums were substituted for the nest of cast-iron tubes originally used for steam and water storage and then steel began replacing cast iron for the generating tubes and for the construction of drums.

Following these developments, the success and widespread acceptance of the inclined straight-tube boiler motivated engineers to experiment with many new concepts and features in boiler design. One such was the invention of Allan Stirling in 1880, a design that connected the steam generating tubes directly to a steam separating drum and provided low head room above the furnace. This became known as the "bent-tube" design, recognized as having certain advantages over the straight-tube design for certain purposes.

Until the latter part of the 19th century, steam was still used largely to provide power and heat for residences of all kinds and local industries. But with the perfection of equipment to provide commercial electric power generation, one of the growing applications of steam was for the ever-widening supply of electricity. Utility companies were formed to take advantage of the power thus provided.

The first commercial electric generating station was designed and built for the Brush Electric Light Company in Philadelphia, which utilized four boilers with a combined rating of about 292 horsepower. This was followed in short order by New York City's first such installation, the Pearl Street Central Station. Thomas A. Edison himself, wearing a white derby, ordered the switch thrown on September 4, 1882, announcing that the opening of the plant ushered in the "age of cities." Boilers like those in use at these two pioneering electric stations were sturdy and reliable for their day. The four installed at the Pearl Street Central Station, for example, were said to have lasted for 20 years, first at Pearl and then at the Edison 53rd Street Station.

It was inevitable that steam power would intrigue ship designers in the early 19th century, first as a means of supplementing the sails of transoceanic vessels that lost money for their owners when becalmed, and then as the only source of power. The *Great Britain,* the first steamship with an iron hull, was an example of the former, having six masts and a full set of sails as well as a steam engine. One of the most successful steamships of her day was *the Britannia,* a wooden side-wheeler of 1,154 tons. In one historic voyage, she embarked from Liverpool for Boston on July 4, 1840 ("defying superstition"), and arrived on July 18. Power for the paddle wheels was supplied by a single-cylinder engine that burned coal at the rate of 38 tons a day and four boilers that operated under nine pounds of pressure. With a rating of 740 horsepower, this power plant drove the ship at an average speed of more than eight knots, quite remarkable for that era.

Such vessels traced their heritage in part to the United States where Robert Fulton had started experimenting with steam at the beginning of the 19th century. As early as 1807, Fulton had placed one of his steamboats in service on the Hudson River, carrying as many as 90 passengers on each trip between Albany and New York City. The ship could attain a speed of almost five knots, which led Fulton to boast on occasion that he passed the conventional sailing vessels that plied the river "as if they had been at anchor."

Steam power particularly captured the attention of marine engineers engaged in the design of warships, which could become sitting ducks when becalmed and relying only on the wind. One of the first successful examples was the *Monitor* of Civil War fame, the invention of a Swedish-born engineer, John Ericsson. Described as "a shingle with a cheese-box on top," this Union vessel fought an historic battle with the Confederate ship, *Virginia* (popularly known as the *Merrimac*). The *Virginia,* also a steamboat, had a record of numerous victories over Union sailing ships, which she could ram at will while they tried vainly to maneuver in light breezes. Both vessels were unique in utilizing propellers, which had an advantage over paddlewheels in that they could not break or become clogged by ice or floating debris. Screw propellers also delivered power more effectively than the paddles.

The marine diesel engine also made its appearance just after the turn of the century, although it was to be many years before it would be commonplace in motor ships. In 1903, some five years after the diesel engine was first produced commercially, this type of power plant was installed in two Russian tankers for limited service in the Caspian Sea. The first significant ocean-going vessel to be diesel-powered was recorded as the *Selandia*, a vessel of 7,500 tons that was launched in 1912. Oil firing became more and more of a prime consideration as labor costs increased. For one thing, oil greatly improved the working conditions of the firemen in the boiler room. For another, it substantially reduced the number of seamen needed. A coal-burning ship the size of *Mauritania*, for instance, required 325 firemen and other laborers just to cart and shovel the ship's consumption of more than 1,000 tons of coal into her boilers each day she was in service.

Even prior to World War I, marine engineers had the option of four different types of power plants which had proved to be practical and commercially competitive. First was the reciprocating steam engine directly connected to the shafts of the propellers, a method that was to exceed all others in popularity for several generations to come. Second was the steam turbine, driving either directly or through a system of reduction gears. Third was the steam turbine driving electric generators which in turn furnished the power to motors that turned the drive shafts. And fourth was the diesel engine, which had still to prove itself but which had some promising advantages in certain types of ships.

Steam also intrigued the inventors of the earliest railroad locomotives, who had failed to harness either animals or the wind successfully in their quest for effective power. In 1825, George Stephenson and his son, Robert, built the first commercial steam locomotive in England for the Stockton and Darlington Railway. It was equipped with a horizontal boiler and two vertical cylinders that were connected to the wheels by piston rods. An improved model, *Locomotion No. 1,* could haul 100 tons of freight and passengers, distributed over 36 small cars, at a speed of about 12 miles per hour. Stephenson was most noted for his *Rocket* which, in a notable demonstration, transported 30 enthralled passengers at speeds up to 30 miles per hour. By this time, Stephenson had perfected his multitubular boiler, vastly increasing the efficiency of his engine and, more precisely, its capacity for making steam. The firebox was surrounded by water compartments, referred to as "water legs," to conserve the heat, a concept that was not essentially improved upon for more than a century.

In 1830, the West Point Iron and Cannon Foundry, located across the Hudson River from the United States Military Academy, produced the first successful locomotive built in America. Designed by an engineer for the South Carolina Railroad, it was aptly named *The Best Friend of Charleston,* for the Southern city from which it would operate. This was also the first all-

17

American steam locomotive to go into commercial service, using coke as fuel and operating at 50 psi boiler pressure.

What was described as "the most successful American locomotive" of its era was built by the Norris Company of Philadelphia. It featured a wagon top boiler, carried atop four driving wheels, which were coupled in order to achieve better traction. The company also built the first locomotive for the Baltimore & Ohio Railroad in 1837 and exported others to Canada, Austria, Germany, and England. Norris was equaled by a later engineer, Thomas Rogers, who used a valve mechanism that permitted the use of steam expansion. Although he went on to produce some 25,000 locomotives, his work was eclipsed by the Baldwin Locomotive Works in Philadelphia, which ultimately became the largest manufacturer of locomotives in North America. The Baldwin locomotive was to become the hallmark of power in all of the major American railroads, as the network of rails stretched across the nation and through thousands of small towns and rural hamlets as well as the great cities.

Noting the increasing success of the steam engine in the development of locomotives, engineers also focused their sights on the road, dreaming of ways to build steam-propelled vehicles that could travel the highways much faster and more comfortably than horse-drawn carriages and wagons. Variations of the steam-driven vehicle came and went, limited in their use and largely objected to and reviled by the populace, which considered them noisy, odorous, and dangerous. They faded from the scene with the advent of the internal-combustion engine. During the decade of the 1860s when the French inventor, Joseph Etienne Lenoir, began producing gas engines that were small and quiet, the internal-combustion engine showed great promise as a commercial success and the steam engine was left to the railroads, the waters, and the realms of industry.

By this time, the cycle of operations for an efficient gas engine had been identified as a simple matter of four strokes: the first to draw in a mixture of gas and air; the second to compress the mixture; the third to ignite the mixture and thereby produce the power stroke; and the fourth to exhaust the byproducts of combustion. With this kind of formula a certainty, who needed all the headaches of trying to harness and control steam in a space that was too small, on a machine that was constantly jolting and braking, and in an environment that was hostile?

CHAPTER 3

BOILER FAILURES
Events Leading Up to the Need for Codes

T hursday, March 2, 1854, was a day Hartford would long remember," began an account in the centennial history of the Hartford Steam Boiler and Inspection Company. "It was sunny and unseasonably warm. People lingered on the streets, seeking excuses to prolong their lunch hours, and availed themselves of open factory, shop, and office doors to drop in for brief chats with friends and acquaintances.

"One such was a young mechanic who had stepped into the engine room of the Fales and Gray Car Works to pass a few words with the operating engineer. At 2:10, while the two were engaged in conversation a short distance from the engineer's post, the boiler exploded with terrific force, completely destroying the boiler room and the adjoining blacksmith shop, and badly shattering the main building. Nine persons were killed outright, twelve died later, and more than 50 others were seriously injured." The investigation that followed the tragedy revealed only that the explosion should never have taken place. The boiler was a new one, having been in use for barely one month. It had been manufactured at the boiler works of woodruff and Beach, one of Connecticut's most reputable industrial concerns and well-experienced in the design and production of commercial boilers. The investigators did learn, however, that shortly before the explosion, several survivors had noted that the steam pressure had risen somewhat above normal. Witnesses testified that the engineer in charge was both competent and diligent. However, several of them hedged a bit, allowing that the man was "a bit careless at times" and also known to have "slipped out occasionally for a beer."

In the end, a coroner's jury concluded that the basic cause of the disaster was "an excessive accumulation of steam" and that "said excessiveness of steam in said boiler was owing to the carelessness and inattention of the engineer."

The jury went beyond the call of duty by volunteering its own observations about what should be done to prevent or at least minimize future accidents. In this, it was unusually perceptive, for the wisdom of its proposals was demonstrated in the following years when many of them were adopted by private or public agencies. Among the proposals were the following:

- Regulations should be devised to prevent careless or inexperienced people from being placed in charge of boilers.
- Regular safety inspections should be made by authorized representatives of municipal or state jurisdictions.
- Boilers should be placed outside of, or blocked off from, the factories for which power is being provided.

- Employers using steam as power should pay close attention to the safety of workers.
- Measures should be adopted to prohibit steam boilers from being rated for higher pressures than would be consistent with safety.

Ten years later, the Connecticut State Legislature did consider one proposal in a positive manner, passing a boiler inspection law. This called for the appointment by no less a personage than the Governor of the State of one steam boiler inspector for each congressional district. Boilers were to be inspected annually and would either receive certificates of safety or be retired from service.

One of the most immediate and fruitful consequences of the Fales and Gray disaster was the founding in 1857 of the Polytechnic Club by a dozen Hartford men who were associated with industries that used steam and who recognized a troubling disparity in industry: although there were thousands of boilers in operation in America, there was an abysmal ignorance of the properties of steam and the causes of explosions. Studying the situation, Club members became familiar with the investigations of two groups in England, the Association for the Prevention of Steam Boiler Explosions and the Boiler Insurance and Steam Power Company whose purpose was not only to inspect boilers but to *insure* policyholders against losses and claims. At that time, the Club seriously considered forming a similar agency and might have done so had not the Civil War disrupted all plans.

How long the boiler problem might have remained in limbo is a moot question, but the next episode in boiler code history rested with a steamboat named the *Sultana*. She was a typical Mississippi side wheeler with two tall stacks. On April 27, 1865, steaming along the river above Memphis, she met a catastrophic end when three of her four boilers exploded for reasons never determined. Ironically, instead of her usual complement of 375 passengers, she was jammed from stem to stern with 2,200 people, most of them Union soldiers who had just been released from Confederate prisons following Lee's surrender at Appomattox. Within 15 minutes, she had burned to the waterline, with a death toll that varied in later reports from 1,200 to more than 1,500, thus providing a grim statistic in the record books as the nation's worst marine disaster in history, past or present.

The impact was astonishingly small on Americans already numbed by the massive tragedies of war — except in the minds of two of the members of the inactive Polytechnic Club, Edward M. Reed and J. M. Allen. They viewed the *Sultana* catastrophe as a dreadful reminder that America needed both public action and a private agency to inspect steam boilers and to insure them.

It was appropriate that such a decision should be made in Connecticut's capital, for the Hartford of the 1860s was already recognized as the "Insurance City of America," as the home of more than a dozen major insurance firms. By the spring of 1866, thanks to the nagging of Reed and Allen, Connecticut's Legislature had before it an act of incorporation for an agency to be called The Hartford Steam Boiler and Inspection Company. With unexpectedly quick approval and an impressive list of incorporators, "The Hartford," as it was to become known, was actively in business. According to Hartford's records, "On February 14,1867, Policy No. 1 was written on three horizontal tubular boilers in the Crompton Loom Works for $5,000, at a premium of $60."

Business was slow in spreading, largely because the whole concept of boiler insurance and inspections was new to industry and few companies wanted to take this unfamiliar step until it had been tried and proven by others. Most of the early policies, issued by company agents, were to factories whose officers were in one way or another connected to the agency. Locating qualified inspectors was equally frustrating since there were no guidelines to go by and little experience in testing procedures. Thus it was that most of the early inspectors were mechanics, boiler shop foremen, or in some cases the agents themselves.

Hartford representatives, travelling around the country to sign up policyholders, ran into an unexpected problem: in some regions where they anticipated doing business, several states were imposing exorbitant taxes on out-of-state insurance companies, an ironic situation in cases where there could be marked improvements in public safety and well-being. Yet within the next decade state and municipal jurisdictions across the United States began accepting company inspections in place of their own as reliable certification that a boiler was in safe operating condition. By the end of the decade, the firm had established a functional engineering department and was offering design services to interested policyholders. The result was such a resounding success that "Hartford Standards" became the hallmark of the boiler industry, incorporating specifications that were adhered to by virtually every major boiler maker in America.

Despite the steadily spreading work and influence of the Hartford Steam Boiler Company, progress toward codes and standards was painfully slow. When research projects were undertaken to provide guidance, they were often inconclusive or short-lived. Typical was the project of a United States government commission which conducted research on boilers between 1873 and 1876. In one project, at Sandy Hook, New Jersey, a team of engineers applied destructive tests to four old steamboat boilers to record their pressure limits. In another project, the team tested five new boilers under steam to determine their weaknesses and recommend practical design improvements. If nothing else, these tests demonstrated that there were great variations in boiler capacities and strengths and that standards were necessary in the industry. The reports also cited the need for regular inspections, both to discover flaws in boilers already in use and to prevent the sale of newly fabricated boilers that were unfit for service.

This was a productive era in the history, not only of boilers and related components, but of engineering materials and products of all kinds. In an engineering article published in October 1987, Russell Berkness, a consulting engineer, described the origins of the American Society of Mechanical Engineers, most notably in regard to ASME's influence at the time on the development of codes and standards in America.

"By 1880, the United States was experiencing a technological transformation," he wrote. "The telephone was a reality and Edison had invented the incandescent lamp. Railroads spanned the continent. In the midst of these developments and the public outcry related to the growing number of boiler explosions, 30 mechanical engineers met in New York City on February 6, 1880, at the offices of the *American Machinist* magazine. Their purpose was to form a society to advance the interest of the mechanical engineering discipline.

"The outcome of this committee's work was the formation of the American Society of Mechanical Engineers. The first president, a prominent engineer and educator, was Robert Henry Thurston, who had taught at the U.S. Naval Academy where steam engineering was established as a major subject.

"Among the first problems to be addressed by the new organization was the development of some procedure for testing the strength of iron and steel, and the establishment of a standard for threads on nuts and bolts. Other standards were necessary to ensure that parts manufactured in one factory and assembled in another would fit properly. Standards had to be identified with scientific precision."

In 1883, the Committee on Standards and Gauges of ASME considered the determination of a standard for rating steam-boiler capability. As one member observed then, "It is part of our duty, no doubt, to establish gauges and standards." In this new spirit of coordination, the Society took a giant step forward when, in 1884, it formulated a code entitled Standard Method for Steam Boiler Trials. One year later, it established a Standard Committee on Pipe and Pipe Threads, thus characterizing the nature of its future work and policies that would be so influential in the later programs to develop boiler codes and standards.

The Centennial History of ASME records the fact that the progression was by no means either smooth or steady. Too many individuals had their own ideas about what should be done, not to mention the numerous critics who felt that nothing at all should be changed or interfered with. ASME's own record of standards became more sophisticated in time, but during that period shortly after the founding of the Society, its history "began with the sort of practical concerns of men who worked with steam engines, boilers, and pumps. William Kent, who proposed that the Society devise 'a standard set of boiler testing rules,' stood directly in the Holley/ Thurston tradition in ASME. He had been a student of Thurston and had worked as his assistant in the strength-of-materials investigation conducted at Stevens Institute for the Iron and Steel Board. In the same way that inadequate knowledge of basic construction materials inhibited their systematic use, the most effective exploitation of steam power required that boiler capability be described in terms that both manufacturers and users understood. One of the problems that prevented the straightforward exchange of information between those two groups was that marketplace competition stimulated extravagant claims for boiler performance, particularly in regard to the amount of fuel required to generate a given amount of steam. But what complicated the matter even further, to use Kent's words, was that 'every engineer who makes a boiler test makes a rule for himself, which may be varied from time to time to suit the convenience or interests of the part for whom the test is made.' To sharpen the message, Kent described his own method of conducting boiler tests, and the ensuing discussion — in just the way he had hoped — made his point that the differences in techniques between them called out for a standard set of procedures."

William Kent, chairman of the first committee to compile a set of rules for boiler testing, pointed out how difficult a job it was going to be to keep such documents current and updated. The 30 or 40 committee members involved with the original code had increased several times over and Kent surmised quite accurately that it would be "a difficult matter to bring them together." A projection of things to come would also have forecast that this would be an ever-increasing challenge for the Society as it grew rapidly in numbers and activities even before the turn of the century.

Shortly after the founding of ASME, other societies and associations came into being that were to make a marked impact on the evolution of boiler codes and standards. One of these was the American Boiler Manufacturers' Association, which was chartered in 1889. Its stated objective was to raise the standards of boiler design and manufacture, and prevent the production and sale of boilers unfit for safe operation. Initially committees were formed on materials, recommended tests and inspections, riveting, tubes, the attachment of valves and fittings, and settings. Three years later, ABMA appointed its first committee on uniform specification laws.

By the turn of the century, the Association was heavily involved in detail work and the presentation of papers on such subjects as riveting, factors of safety, caulking, dished heads, flanging, tubes, bending and forming, stay bolts, braces, drums, and hydrostatic-pressure tests.

Colonel Edward D. Meier, who was to figure prominently in the birth of the ASME Boiler Code was quoted in his position as chairman of the ABMA Uniform Specification Committee as being dissatisfied with one direction that committee work was taking. Too many members were looking at specifications in light of bidding for boilers and components and not toward the adoption of standards and codes by states and municipalities. This outlook, he felt, was way off balance. The concepts he adhered to during this period in his career reflect the ideas he would later bring to the creation and development of the first ASME Code.

By the end of the century, there still were very few localities in the country that had paid much attention to the enactment of laws and regulations for boilers, despite the continuing number of explosions, casualties, and property damage that were laid at the feet of faulty worksmanship, sloppy installation, or inept supervision by the users. The city of Detroit was one of the few exceptions, having enacted in 1869 a fairly effective ordinance relating to the inspection and care of steam boilers and their operation.

Following the formulation of A Standard Method for Steam Boiler Trials in 1884 under the direction of William Kent, ASME published a number of papers and reports that would later assist the Boiler Committee in compiling background materials for writing the first Code. One example was The Standard Method of Conducting Duty Trials of Steam Pumping Engines, published in 1891, the work of a five-man committee. The year before that, another committee had devised a standard method of conducting locomotive efficiency tests, which later was split into two parts, the first to cover shop tests and the second to cover road tests.

Despite these and other efforts at the end of the 19th century, there was still no legal code for safe stationary boilers in any of the states of the Union. Massachusetts had considered the enactment of laws and regulations because of the prevalence of steam boilers in hundreds of factories and mills throughout this increasingly industrialized state, but legislators had become complacent for two reasons. The first was the positive influence of Hartford Steam Boiler Inspection and Insurance Company throughout New England, whose agents had done an excellent job of policing equipment which the firm insured. The second was that between 1898 and 1902 there had been no serious boiler explosions reported in any of the industrial regions of the state, a dramatic contrast to the nation's total of more than 1,600 during the same period. This was attributed in part to the fact that Massachusetts had passed a law in 1850 requiring fusible plugs on all stationary high-pressure boilers.

This complacency was shattered quite abruptly on March 10, 1905, when a fire-tube boiler in a Brockton, Massachusetts, shoe factory exploded. The toll was 58 dead, 117 injured, and damages of one quarter of a million dollars. Since the factory boilers were not insured and the cause was never officially determined, it was evident that the state laws were not as effective as had been claimed. Emotions ran high during the proceedings of the State Legislature which was then in session in Boston, especially when one legislator charged that there *had* been boiler explosions all along, but that manufacturers had silenced such reports, fearing that regulation of boilers would add heavily to operational costs. The issue quieted down and the Legislature did not take any immediate action. On December 6, 1906, however, another serious explosion took place at a shoe factory, this time in Lynn. Although only one person was reported killed, this incident motivated the Governor of Massachusetts to include in his inaugural address a month later a demand for prompt action.

The wheels were set in motion, a five-man Board of Boiler Rules was authorized that spring, and by the late summer of 1907 the first Massachusetts Rules were approved. The document was short and simple, containing only three pages. The first was devoted to a facsimile of the standard format of the certificate of insurance. The second page covered fusible plugs and their performance characteristics, based on the earlier state requirements for these safety devices. The third page provided specific rules, which included among others limiting cast-iron boilers to a pressure of 25 psi, limiting boilers with cast-iron headers to 160 psi, and data governing the shearing strength of rivets.

Enactment of the Massachusetts Rules was not without dissension, most of the objections coming from manufacturers who viewed these regulations as a prime example of needless government intervention. Some denounced the state for imposing commercial hardships that would put small boiler makers out of business. The hue and cry, along with legislative lobbying, forced a public hearing in 1909 to listen to complaints and recommendations for revisions. Attending this hearing was Dr. David S. Jacobus, who had arrived from New York, representing the Babcock & Wilcox Company. Dr. Jacobus was later to be recognized for his fine work on the ASME Boiler Code, but in this instance he was branded an "outsider" and criticized for coming all that way to make proposals that would be injurious to the welfare of manufacturers in Massachusetts. Jacobus responded by stating that it was his personal policy, as well as that of his company, "to act in the broadest way possible to endorse a movement for the protection of human life and property."

Following his statement and leadership, others at the hearing spoke out in favor of the Rules. The result was that "An Act Relating to the Operation and Inspection of Steam Boilers" was passed in 1909. The rules were divided into three parts. The first applied to boilers installed prior to January 1, 1909, fixing the maximum allowable pressures for boilers composed of steel and wrought iron. It also specified the sizes of non-spring-loaded safety valves and bottom blow-off valves. Part two referred to boilers installed "now and in the future," defining maximum pressures for cast-iron boilers, for boilers with cast- or malleable-iron headers or with cast-iron mud drums. Part three covered boilers of the future, anticipating requirements for materials to be employed in the fabrication of various components. It also described the procedures for stamping boilers that met the requirements of the rules and provided guidelines for every kind of component, as well as non-standard boilers and portable boilers. The document concluded with an appendix devoted to structural recommendations and the care and operation of boilers in service.

The success of the Massachussets law, along with public pressure to do something about continuing boiler explosions, motivated another state, Ohio, to take similar action. In October 1911, the Governor approved the Rules that had been formulated during the previous five months by the Ohio Board of Boiler Rules, which had been appointed for that purpose. The Ohio Board adopted with few modifications the Rules of the Massachusetts Board, in most cases, changing only the dates to refer to boilers constructed prior to, and after, the passage of the bill.

It had taken 80 years for the pioneering concepts of various Code advocates to take root in America. Yet this was just the beginning. Even as groups of civic-minded engineers in Massachusetts and Ohio were actively formulating their codes, another group within the ranks of the American Society of Mechanical Engineers was looking to the future in a way that would change forever the entire boiler industry and its future evolution.

CHAPTER 4

PERSONALITIES AND PIONEERS

The People Who Created the First Code

One of those appointed in 1911 to compose the initial draft of the Boiler Code was Colonel Edward Daniel Meier, who had studied in Germany, where he graduated from the Royal Polytechnic College in Hanover before returning to the United States. After distinguished service with the Army of the Potomac during the Civil War, and a rise in the ranks to the position of Colonel, he designed machinery for compressing cotton and in 1884 was one of the founders, and later president, of the Heine Safety Boiler Company. He designed and installed boilers in New York City's new Grand Central Station and was the first to introduce the Diesel engine to the United States, after it was patented in 1892.

At the time of the formation of the new Boiler Code Committee, he was president of ASME and also of the American Boiler Manufacturers' Association. A distinguished looking man with snow white hair and walrus mustache, he had the potential to be a driving force in the creation of the Code and was, in fact, a motivating force in getting the project underway. Some 12 years earlier, he had attempted without success to motivate the American Boiler Manufacturers' Association to consolidate a voluntary set of rules for the construction of steam boilers that would be uniform across the country. Now he decided to try to attain the same objective within the framework of ASME, hoping that the technological climate would by this time be less hostile.

As early as 1898, he had presided over a committee whose objective was to standardize specifications for the construction of boilers. He was ably qualified for that assignment, having already made an exhaustive study of European standards and surveyed the practices and policies of major American boiler manufacturers. But the undertaking was ahead of its time, weakened by the constant harping of critics who felt that the restrictions imposed would pose economic and competitive threats to the boiler industry.

Meier's first step was to request that the ASME Council appoint a committee to formulate standard specifications for the construction of steam boilers and other pressure vessels and their components. Ironically, while in the midst of trying to achieve his goal of industry-wide standardization, the Colonel became ill and barely survived long enough to see the final version of the Code approved for publication.

Fortunately, there was a fellow engineer who was an equally strong influence in creating the Code: John A. Stevens, of Massachusetts, a consultant whose career had been devoted largely to power generation. He had already amassed a wealth of experience in the area of codes and standards, having served on the Massachusetts Board of Boiler Rules, which as early as 1907 had established the first state law to regulate the construction, installation, and operation of steam boilers. Significantly, the Board was not simply a state regulatory body, but was composed of individuals representing boiler manufacturers, operators, inspectors, and owners. This varied structure tied in with a basic objective of the first ASME Code: to assure complete and well-balanced representation on the part of those who were concerned with consistency, efficiency, and safety.

When the ASME Council appointed Stevens chairman of the Boiler Code Committee in September, 1911, it was virtually assured that this kind of balance would be established and maintained. "In essence," wrote Bruce Sinclair in his history of the Society, "ASME's boiler code adopted the Massachusetts rules almost wholesale and then established a process for criticism and revision by the technical community." Basically, the only real changes involved the reduction of allowable steam pressure in certain cases and the addition of information about design.

The Boiler Code Committee, as approved by the ASME Council, was listed in the minutes of the meeting of September 15, 1911, as follows:

"Voted to confirm the appointment of a committee to formulate standard specification for the construction of steam boilers and other pressure vessels and for the care of same in service to consist of John A. Stevens, chairman; E. F. Miller, C. L. Huston, H. C. Meinholz, Richard Hammond, R. C. Carpenter, and W. H. Boehm."

Following the policy of diversity initiated by the Massachusetts Board of Boiler Rules, the new committee represented a balance of interests. Stevens, as has already been mentioned, was a consulting engineer in the field of boilers. William H. Boehm was an insurance engineer and at the time employed as an officer of the Fidelity and Casualty Company of New York; Professor Edward F. Miller and Professor Rolla C. Carpenter were educators; Charles L. Huston was a producer of iron and steel; and H. C. Meinholz and Richard Hammond were both boiler manufacturers.

An important addition to this group was C. W. Obert, who was hired as the first official secretary and who has been credited with shouldering a great deal of the administrative work of the Committee during 12 years of service in that post. He was not to become an appointed member of the Committee until he resigned as secretary in 1927.

Another key advisor was William Kent, mentioned earlier as a Canadian with long and varied experience as a mechanical engineer including positions as editor of *Iron World* and *American Manufacturer,* manager of the Pittsburgh office of Babcock & Wilcox, and president of the American Society of Heating and Ventilating Engineers. With William Zimmerman, he had founded the Pittsburgh Testing Laboratory and was later heavily involved in the investigation of steam boilers and the combustion of fuels. Greatly concerned about the diversity of standards and the absence of coordination in his field, he set up presentations to demonstrate to others in his profession, particularly mechanical engineers, that in the matter of testing steam boilers alone, each person viewing the demonstration was probably using methods that differed from those applied by others in the audience.

The point was well taken, particularly when another engineer, Henry R. Towne, complained to all who would listen that there was not even a common language with which engineers could communicate many of the differences.

The first two years of the Code Committee were devoted to organizational work and two meetings, the first in New York City, the second in Ithaca, New York. Then, in 1913, following procedures that were already well established in the American Society of Mechanical Engineers, Obert typed up the Preliminary Report of the Committee and sent some 2,000 copies to the Society's list of interested professionals. These included educators in the field of mechanical engineering; inspection superintendents at casualty insurance companies; chief inspectors of national, state, and municipal boiler inspection agencies; engineers interested in the construction and the operation of steam boilers; manufacturers of boilers; and editors of engineering journals.

The report was essentially the Code itself, requiring 230 printed pages and including charts, tables, laws, rules, and appendices.

A covering letter requested that the recipients of the report study it carefully and recommend additions, deletions, or any other revisions they deemed to be suitable. The end purpose, of course, was to draft a Code that would be approved across the board by all who were affected by it and that would reflect the varied interests of everyone in, or affiliated with, the boiler industry.

Following the receipt of suggestions that were triggered by this mailing, the Committee eventually completed revisions for a second printing, dated February 18, 1914. Comparison of the two editions shows relatively few changes of significance, which would seem to indicate that few of the recipients had been dissatisfied with the initial effort. As in the previous case, copies of the new printing were then mailed to the ASME list, now increased to some 2,500.

Explaining the plan and procedures it was following, the Committee invited all recipients to "cut out what you do not consider advisable to incorporate in the final copy of the Report and also to add to our work whatever will, in your judgment, further safeguard human life and property."

The Committee went out of its way to solicit critiques, obviously concerned that the percentage of responses on the first go-round had been lower than anticipated. "It will be inferred," warned the accompanying letter, "that no criticism from you means that you approve the publication of the pamphlet substantially as sent you with the additions which may come from conference with the members of our mailing list."

"You will observe," said the closing statement of the letter, "that we are trying to give everyone qualified an opportunity to present his opinions, so that the rules may represent the best judgment of a majority of the qualified engineers of the world."

Unlike the first mailing, the second one brought what was later termed "a storm of protests," particularly during the month preceding the Spring Meeting of the Society, which had been scheduled for mid-June in Minnesota's twin cities of Minneapolis and St. Paul. Although dissension had not really surfaced until now, it was apparent that there had long been dissatisfaction among members of ASME who were opposed to rules and restrictions governing their businesses. In the proposed Boiler Code they had seen a threat to their economy and even to the whole free enterprise system. There was a movement underfoot to scuttle the issuance of the Code by planting seeds of doubt in the minds of the Council members themselves so that the publication would be voted down, or at least postponed.

Why had not the first mailing stirred up this hornet's nest, and why was it only at this late hour that opposition was mounting? As one opponent admitted, "I guess we all thought that if we ignored the Code it would simply go away."

A more logical conclusion is that many who opposed the Code did so because they felt that the Committee was too heavy-handed, if not autocratic. It had been given a unique kind of power and now was trying to dictate the rules. Some laid the blame squarely at the feet of chairman John Stevens, who was described as having an "assertive personality" and who openly considered himself an authority on the subject because of his past work with the Massachusetts Board. His very association with a governmental agency, however, was damning, at least in the eyes of ASME members who had always devised policies and taken action on behalf of the private sector rather than the government. It was claimed that the Committee had worked "independently," rather than following the ASME principles of widespread industry participation.

So tense had the situation become that on Monday, June 15, two days before the Spring Meeting was to commence, the Boiler Code Committee met in Chicago with representatives of industry associations to try to enlist support for the Code. From the moment the Spring Meeting opened, however, on June 17, it was evident that opposing forces were lined up for battle. One leader of the opposition was Henry Hess, a member of the ASME Council, who severely criticized the Preliminary Report and personally recommended that the Committee be "reconstructed," that the Report be discarded, and that the creation of the Code be started all over again from scratch.

It was even suggested that Stevens be removed from his position as chairman. For a time, the members of the Committee were in shock. The work that they had devoted so much time to over the past three years was to be discarded like an old shoe that did not fit. The day was saved, however, when several supporters objected to the Hess plan. One of these was Dr. Jacobus, who had continuously maintained his interest in codes and standards. Much of his experience had been gained in original investigation and research in the study of mechanical devices and processes for the production of power and ongoing efficiency tests of steam engines, turbines, and other power plant apparatus. Jacobus emphasized that the Committee had maintained the highest standards in putting the proposed Code on paper and that John Stevens, instead of being castigated for being too aggressive, should be honored for the role he had played in formulating the Massachusetts Rules on which the ASME Boiler Code was based.

As a man who was described as one who "knew how to deal with people," Jacobus was equally persuasive in calling attention to the fact that Edward D. Meier, whose brainchild the Code had been originally, was very ill and that it would be a great blow to him to learn that what could be his most significant contribution to the Society was to be obliterated. He advised that a public meeting be held to assure widespread industry participation in the delineation of the Code.

In the end, both sides won. Hess and the others who had expressed strong opposition backed off and agreed that the Committee should be allowed to continue its work. For its part, the members of the Committee conceded that they would make every effort to review the proposed Code with representatives of all factions involved and make suitable revisions before publishing the official First Edition. To that end, a Society resolution was passed, calling for a public hearing to be held on September 15, 1914, at which time a review would be made of all suggestions, criticisms, and reports submitted in writing on or before August 15.

It was evident that the Committee took this new step with great diligence. Almost 40 organizations were invited to send representatives, making certain that no interested group would be bypassed. The public hearings were held at the Engineering Societies Building at 29 West 39th Street in New York City on September 15, as scheduled. The 150 individuals who attended represented every facet of the industry, from consulting engineers, educators, and editors, to manufacturers, insurers, inspectors, government officials, agriculturists, railroaders, designers, heating and energy specialists, and researchers. The list of associations present read like a blue book of the industry.

Unlike the tone of the Spring Meeting, the atmosphere at this fall conference was one of anticipation and cooperation. The participants and Committee alike expressed confidence that differences would be resolved and that the Code would become a living instrument that would work to the benefit of all parties. John Stevens addressed the participants in a very positive voice, pinpointing the factors, such as safety, uniformity, and efficiency, which made the Code an absolute necessity. It was evident, however, that he had been somewhat coached or at least advised by others who were more moderate because his presentation was restrained enough so that he did not irritate his former detractors.

At the conference, after preliminary discussions, the nature and wording of the Code were discussed item by item so that everyone attending could express opinions whenever changes had to be undertaken. The accomplishments were notable. It was noted, for example, that "for the first time in their history" all of the makers of safety valves agreed upon a uniform specification for their products. And representatives of the railroad industry presented "a most splendid criticism of the Preliminary Report, which helped greatly in bringing about the actual success of the hearing."

One of the most meaningful discussions pointed out the need for additional studies of particular issues, creating appropriate subcommittees, and establishing a workable process for continuing revisions automatically in years to come. Thus it was that the creation of the first edition of the Code also served as the prototype for the procedures and systems that have been carried down by ASME to the present day.

The official transcript of the hearings led immediately to preparations for what was actually the third draft of the Code, though still referred to as the "Preliminary Report." Incorporating the critiques compiled during the hearings and working around the clock, the Committee managed to have the draft ready to send out to its mailing list by November 5, a gargantuan accomplishment in light of the amount of detail that had to be covered with meticulous care. Even more challenging was the deadline for a *fourth* printing, which had to incorporate changes requested on the third printing and be ready for presentation to the ASME Council for its Annual Meeting on December 1.

The "Introduction" to the four Reports, which changed very little from one draft to the next, is significant because it reveals the philosophies and outlooks of the founding fathers of the Code. After presenting an historical synopsis of boiler explosions, damages, and casualties, the text made it clear that such disasters called for every effort to design, construct, and install steam vessels and their appurtenances in "as nearly perfect a manner as possible."

The introduction then emphasized the intolerability of a situation whereby the laws and restrictions from one area to another were such that the interstate shipment of units and parts was almost impossible without breaking existing laws. It was firmly stated that this "intolerable confusion" seriously affected "virtually every manufacturing interest in the United States." The message was clear to all recipients: it was far more favorable to their interests, as well as less restrictive in the end, to accept a uniform Code than to continue without one.

The lengthy introduction summarized the sequence of events during the preceding four years, from the formation of the Boiler Code Committee to its well balanced makeup, the effort to submit reports to 2,000 and more engineers across the land, and the repeated invitations to all those interested to study the Preliminary Report and respond with their considered critiques. Stressed also was the fact that, at the hearings, *agreements were reached* on uniform specifications for all of the major categories, including boiler steel and other basic materials, boiler tubes, safety valves, fire-tube boilers, water-tube boilers, steam and hot water heating boilers, and other products and components.

It would seem from all outward appearances that a spirit of cooperation had replaced the notes of dissension that previously delayed the acceptance of the new Code. But at the meetings held during the first week of December 1914, there was a new campaign on the part of the opposition to defeat the approval of the Code, or at least to postpone the action. However, the supporters now greatly outnumbered the dissenters and had so firmly entrenched themselves that the Committee was able to proceed on the final draft of the Code. According to the records of the secretary, C. W. Obert, the seven-member committee and the 18-member advisory committee labored 13 hours a day, six days a week, for seven weeks to incorporate the changes and additions that had been presented to them. The galley proofs for the Code were presented to the Council at its regular meeting on February 15, 1915. In its published form, including a detailed 30-page index, the Code was 148 pages long. By the time of the next Council meeting, on March 12, 1915, the Code had become an official, approved document of the American Society of Mechanical Engineers. Despite being actually published in 1915, the first Code went on record as the *1914* Edition, which thus became the official birthdate of the Code. The format was a hardcover book, 6" x 9" in size, and bound in olive drab cloth. Entitled REPORT OF THE BOILER CODE COMMITTEE OF THE AMERICAN SOCIETY OF MECHANICAL ENGINEERS, it was published at the expense of Babcock & Wilcox.

In the preface, the Committee described the document as its "final report on Rules for the construction and allowable working pressures of stationary boilers, which forms a portion of the task assigned to it." It was emphasized that the primary objective in researching and publishing the Rules was to secure safe boilers without placing undue hardship or demands on boiler manufacturers, installers, and users. Looking toward the future, the members recommended that a permanent Committee be established to make revisions as needed, at least once every two years, and "to modify them as the state of the art advances."

It is interesting to note here that the term, "state of the art," which is considered such a modern designation, was part of the ASME idiom more than 75 years ago!

This first Code book was divided into two parts, PART I on new installations and PART II on existing installations. Realizing full well that at this stage in the formulation of Rules, it would be difficult to change what already existed, the Committee devoted 80 pages to new installations and only five to those already existing. The remainder of the First Edition was allocated to an Appendix, with suitable diagrams, formulae, and charts, and a detailed Index. Power boilers and heating boilers were both covered in the main body of the text, with particular attention given to such topics as the selection of materials, manufacture, the thicknesses of the components, workmanship, inspection and testing, safety valves, water and steam gauges, fittings and appliances, and official ASME Stamps for uniform standards.

It was evident from the results that the pioneering Committee that had been appointed in 1911 had done its homework so well that the First Edition was a thorough and dependable prototype for editions that would follow. Thus it was that, even four years later, the Edition of 1918 was in some respects almost a carbon copy of its predecessor. That is not to say that there were few revisions or modifications, but rather that they reflected the changing "state of the art" and not fundamental reconstructions of the Rules and related data.

The most significant recognition was that of special considerations necessary for boilers of the forced-circulation or flash type, for which new matter was incorporated. Another important addition was a preface devoted in large part to explaining procedures for handling outside inquiries about applications of the Rules and investigating any matters of controversy or criticism. To this end, the Committee had taken positive steps toward encouraging the participation of representatives of states and municipalities, to attend meetings, help with interpretations of the Code, and otherwise cooperate in expediting revisions and updating. This attitude of coordination and cooperation had manifested itself well before publication of the first Code and notably in mid-September 1914, when a conference was attended by the Committee and representatives of a number of organizations active in the industry, including the American Boiler Manufacturers' Association, the Association of American Steel Manufacturers, the National Tubular Boiler Manufacturers Association, the National Association of Thresher Manufacturers, and leading producers of water-tube boilers. At that time, approvals had been given by these groups for the specifications that had been drawn up in the Code for boiler plate steel. In addition, specifications for lapwelded and seamless boiler tubes had been approved by the Boiler Tube Manufacturers of America in the early fall of that historic year.

Although *safety* has popularly been cited in histories relating to the creation of boiler codes as *the* motivating force for action, there were two other factors that directly prompted ASME to take action. One was the *economic benefit* that would be derived from the avoidance of waste and unnecessary work trying to join components that had never been designed to match each other. The other influence was *professional honor* and the reputation for communication and coordination.

Speaking about the early attempts at standardization by ASME groups, the Society's official history itself supported the concept that no action in its past reflected those three impulses more graphically than the effort to reduce the number of steam-boiler explosions through standardized procedures for their construction, use, and inspection. As was emphasized, too, the steam-boiler industry was in need of rationalization since both the producers and the users of boilers and components were constantly frustrated by a "crazy quilt of regulations" governing their design, construction, installation, and use. At the time the Committee was formed in 1911, there already were numerous laws and regulations in effect, most specifically in ten states and 19 metropolitan areas. Specifications were so varied, however, that it was extravagant from a financial standpoint and counterproductive from an engineering standpoint to install equipment approved in one area to a different location where the certifications would be invalid.

It was true, of course, that prior to publishing its own Code ASME had hailed the developments in Massachusetts and Ohio as progressive steps that would do much to improve public safety and enhance the image of an industry that was constantly coming under fire because of accidents, casualties, and breakdowns. Yet few members who were knowledgeable about boilers had been lulled into believing that these states would become bellwethers for a flock of other municipalities and states to follow. To the contrary, many believed that most jurisdictions would probably adopt a "wait-and-see" attitude and delay legislative action until some pronounced statistics were on the board.

ASME had recognized, too, that while the rules adopted by Massachusetts and Ohio were appropriate at the time for the locations and circumstances concerned, they were more "regional" than "national" in scope. One of the major problems that had long existed was the lack of *uniformity* that placed manufacturers in the awkward — and often financially unwelcome — position of having to modify boilers before they were shipped in order to conform to different regional requirements and traditions. A major manufacturer who desired to conduct business in a dozen states, for example, had to carry a prohibitively large inventory of components and parts to meet the requirements.

Added to this problem had always been the frustrating and trying matter of inspection. The laws were so varied from region to region, even from city to nearby city, that many a substantial boiler order had to be rejected because the manufacturer had not interpreted the local inspection laws accurately. It was embarrassing, professionally as well as financially, to have a large boiler given a black mark because certain fittings fabricated in New York happened to be somewhat different than the ones acceptable in Pennsylvania.

By 1911, the situation had become what was described by one disgruntled manufacturer as "more confused than a cat with a rubber mouse." So it was with enormous relief — even to many manufacturers who had protested the actions inaugurated that year by the Boiler Committee — that the Code finally became a living reality.

Despite the thoroughness of its initial effort and the extensive amount of time and thought given to the movement by the pioneers who participated in it, the first Boiler Code Committee committed one fundamental mistake. It set its sights on perfection, yet at the same time propelling itself into the publication of the first edition of the Code in order to meet the deadlines it had assumed. Some critics referred to this action as "technological arrogance," motivated by a belief that it could do no wrong and that earlier, essentially unsuccessful, attempts by manufacturing groups to establish codes failed because the participants were looking at the challenges from a commercial viewpoint rather than the lofty plateau of a learned society.

Only time would prove whether the Committee or the critics had the gift of foresight.

CHAPTER 5

BUILDING
A HEAD OF STEAM
All Work and No Relief for the Committee

One of our primary objectives has to be to talk down the dissenters, many of them right within our own ranks, and convince them of the great common good that will eventually come from the Boiler Code. That must be high up on our agenda, for without uniform agreement and coordinated action, the very purpose of our work will be undermined."

Speaking at a hearing shortly after the publication of the first Code, Francis Winthrop Dean voiced the feelings and concerns of fellow members of the 18 members of the Advisory Committee. Dean, who had received his bachelor degree from Harvard *magna cum laude* in 1875, and later his M.E., had been a designer of industrial pumps, engines, and boilers. As a consulting engineer, he specialized in both stationary and locomotive steam power problems, internally fired boilers of large capacity, high-pressure steam, and vertical engines. As an ASME Vice President, he was considered a senior spokesman on many of the Society's major issues.

Partly because of the public relations campaign he vigorously supported on behalf of stronger coordination, there was less of a split in the ranks than many had first feared. Opposition to the new Code faded as more people in the industry came to realize the need for both safety measures and uniformity. A notable exception was John Clinton Parker, a Philadelphia engineer engaged in the manufacture of boilers who made no bones about his dislike of regulations. He was outspoken in his resentment of ASME for instituting the new Code, as he asserted in a characteristic letter to the Council in 1914 when he learned that a document was being drafted that would make him adhere to certain rules.

"The writer desires to register a strong protest against further backing of the propaganda for state control of boiler design, with the funds and at the meetings and in the publications of the Society." He accused the Committee of devious and underhanded dealings as an illegal way of trying to "sabotage" competitors such as his own company, and he did not draw the line at using outright slander and exaggerated claims to make his point.

At first, the Council and the Committee made every effort to be fair and to listen to his complaints. But it soon became apparent that Parker would not be placated. Instead, he not only continued his rantings but went so far as to place unauthorized Code Symbol Stamps on his company boilers before shipping them to customers. When ASME, as it was obliged to do, informed state authorities about this violation of the law, Parker charged "commercial discrimination" and made it clear that he was not through either with his tirades or his campaign to discredit the Society.

"It may be difficult for some younger engineers to believe that opposition to, or doubts about, the success of a boiler code could exist or that any spirit except that of greatest enthusiasm could greet proposals to eliminate the chaotic conditions that existed prior to the adoption of the ASME Boiler Code."

This was the theme for an editorial in the July 1952 issue of *Mechanical Engineering,* which eulogized ASME pioneers who refused to give up, despite "the long and arduous campaign, some of it carried on in an atmosphere of skepticism as to its probable success." The editorial stated that the establishment of the Code had required almost four decades of continuous work by hundreds of engineers before stifling the critics and had earned widespread acceptance by insurance companies, users, manufacturers, and government agencies.

"The task was not an easy one," concluded the editorial. From time to time crises arose which might have spelled defeat. But the system was a good one and the devoted men who administered it were patient and wise. Today the Code stands as an impressive example of the successful working of the American system of free enterprise at its best.

Although the Parker protest subsided and the Committee continued its work of continually attempting to improve and update rules and specifications, it was evident that an undertaking of this magnitude could never be as "perfect" as had been idealized in the preface to the First Edition. What might seem right on target for one manufacturer or user could not always prove to be right for all others affected by the Code. This fact of life was recognized then and has been throughout ensuing years. An analysis of the situation was presented by two ASME members, J. R. Farr and J. F. Harvey at a pressure vessel and piping conference in Chicago in 1986 in a paper entitled, "Why So Many Different ASME Code Formulas for a Cylindrical Vessel?"

"At first thought," they wrote, "it certainly would seem that the ASME Boiler and Pressure Vessel Codes would contain only one formula, rule, or criteria that would apply to the safe construction of a simple right cylinder subjected to internal pressures. Not so! The ASME Codes for boiler and pressure vessels did not always proceed in a preplanned and orderly manner."

They cited the first code formula as:

$$P = \frac{TS \times t \times E}{R \times FS}$$

P stood for the maximum allowable working pressure in pounds per square inch; *TS* the ultimate tensile strength in pounds per square inch; *t* the minimum thickness of shell plate; *E* the efficiency of longitudinal joints or ligaments between tube holes; *R* the inside radius; and FS the factor of safety.

"This formula simply gives the permissible pressure based on an allowable average membrane hoop stress in the vessel wall thickness. There was nothing wrong with this approach. Pressures and temperatures were low and shells were thin compared with their diameter. Boiler and pressure vessel shops were numerous and few had engineers in their employ; hence, some degree of sophistication had to be sacrificed for the safety that accrues from simplicity."

It was interesting to note that allowable external tube pressures that had been established experimentally some three quarters of a century ago were extensively reinvestigated in 1983 and found to be "completely reliable."

Few engineering feats in modern times can compare with this endurance record. Instead of the concept advocated by so many industries — "planned obsolescence" — this was the case of the opposite — "perpetual life."

Citing the accomplishments of the Code as "remarkable," Farr and Harvey went on to relate that, "for the next decade or so, there was little change in this code formula for a cylindrical vessel. There was little need for change. Vessels were of low pressure and mostly of riveted construction (seldom exceeding 285 psi) with thin walls; and most of all the safety record continued to be most impressive."

From that point on, however, "it was inevitable that slightly different formulas for the same straight geometric cylinder, subject to internal pressure, dependent upon their diameter and thickness, would arise."

Trying to effect a marriage of *accuracy* and *simplicity* became increasingly difficult and something of a paradox. In the endeavor to use the increasing storehouse of knowledge, simplicity was necessarily lost. Also, as more and more designers, manufacturers, and contractors entered the scene, the Code continually faced the challenge of preserving construction safety for all vessels made by all fabricators.

"As the ASME Boiler Code comes into use more and more throughout the country, questions are put to the Society regarding the meaning or application of particular rules embodied in it." This statement in the Committee Transactions for 1916 added that during the year 112 inquiries had been considered and properly answered during monthly meetings. A further indication of widespread Code recognition was that it was being used in technical schools as a text or reference work, notably at Stevens Institute of Technology, the Sheffield Scientific School at Yale, Rensselaer Polytechnic Institute, Tulane, Georgia School of Technology, the University of Texas, and Virginia Polytechnic Institute. Furthermore, the Code had been adopted by 15 of the leading boiler insurance companies.

From its inception, the Boiler and Pressure Vessel Committee had started to open up avenues of communication and public participation. One of these efforts resulted in the policy of publishing revisions to the Code (or abstracts) in the pages of *Mechanical Engineering.* Another avenue was continuation of the policy that had been adopted even before the first Code was published: scheduling public hearings to discuss proposals for new sections in the Code, as well as major revisions. Comments and suggestions that resulted from these contacts were then referred back to the committee responsible for the subject under discussion.

In addition to these early established pipelines to assure a more balanced Code, the ASME Council took an important step in 1916, when on October 13 of that year it approved the formation of a Conference Committee to be composed solely of representatives appointed by states or municipalities that had adopted, or would be adopting, the Code. Referring to these conference groups, the preface stated, "Since the primary function of the Code is to promote the safety of the public, the Boiler and Pressure Vessel Committee seeks the cooperation and guidance of the public by means of conference groups set forth hereunder."

Describing why it was that these conferences inevitably ended up wrestling with so many technical details, retiree Frank S. G. Williams pointed to flanged outlets as an example. Williams, who had begun service in boiler and pressure vessel work in the 1920s and was later chairman of the Boiler and Pressure Vessel Code Committee, explained, "The original scope of the power boiler Code ended at the threaded or flanged outlet from the boiler drum. By the mid-1920s, the scope had been extended to the first valve in the steam piping and later to the second stop valve, as the batteries of the boilers became more and more commonplace."

Why did this expansion take place? "In the period immediately following the adoption and widespread use of the Boiler Code," he said, "several fatal accidents occurred that were puzzling at first. These took place both on land and aboard ship during the hydrostatic test and during the startup testing of boilers. I can still recall the disfigured face of the Chief Civilian Engineer of the United States Steamboat Inspection Service. He had been the victim of one such accident and, though he recovered, bore the scars of that boiler accident for the rest of his life. Since he was active on Code committees, his scarred face was a constant reminder that the rules still had to be improved."

In addition to the Chief, Williams recalled that at least two inspectors had been seriously injured and one was killed during that era in a similar explosion. Because of these casualties, it was decided to shield that portion of the piping which was going to be tested so that personnel associated with the testing of boiler installations would be protected.

Walter B. Parker, who served as Chairman of the Uniform Boiler and Pressure Vessel Laws Society, spoke out often about the number of serious boiler accidents and casualties. He cautioned engineers, as well as the public, that the act of publishing codes did not automatically guarantee safety. "Some people," he stated, "are reluctant to observe laws relating to boilers and safety. They did not — and do not — *understand* boilers, their operation, and particularly the hazards that are always imminent, being mainly interested in not having any government regulation of their business. Many of these people honestly accept explosions as 'an act of God' that could not be prevented.

"So now we have the enigma of having a well written and Council-approved Boiler Construction Code, which has no standing whatsoever until it has been legally adopted by a jurisdiction. How could the jurisdictions be influenced to adopt the Code as their construction and installation requirements?"

At one of the later sessions of the public hearings on the 1914 Code, the question had arisen about the practicality of forming an organization to bring the Code before various legislatures. At the time, the American Boiler Manufacturers' Association had a Committee on Uniform Boiler Laws. A member of that committee, Thomas E. Durban, realized the need not only for uniformity in the construction codes, about which he had been addressing manufacturers associations, boiler users, and government officials for several years, but for rules and regulations for inspection that would be *based on laws adopted by the jurisdictions.*

As a direct result of Durban's interest and campaigning, the American Uniform Boiler Law Society was formed early in 1915. Its primary objective was to present the ASME Code to governing bodies of states, provinces, and cities so as to secure legal adoption of the rules. Durban was appointed chairman of the AUBLS in 1916, at which time it had a Council of 12 men who represented power boiler manufacturers, heating boiler manufacturers, pressure vessel manufacturers, railroads, public utilities, and the producers of boiler equipment and appliances. In this manner the AUBLS was formed, later incorporated under the laws of New Jersey, and then in 1954 changed its name to the Uniform Boiler and Pressure Vessel Laws Society to indicate its broader coverage.

From the start, the Society made it clear that it could function property only if it had well-balanced representation and if membership were open to anyone who wished to contribute to its support and promote its objectives.

"What are the objectives of the Society?" asked Walter Parker when, at a later date, he was appointed Chairman. "The Society is a non-political, nonprofit, technical body that is supported by the voluntary contributions of its membership. The Society believes that any laws, rules, or regulations should follow nationally accepted codes and standards."

Following through on this policy, Thomas Durban was instrumental in planning proceedings for representatives of government and industry which resulted in the American Uniform Boiler Code Congress. Held in Washington in December 1916, the Congress was attended by participants selected by the governors of 22 states and one Canadian province, the government of the District of Columbia, and the mayors of four cities. A report of the sessions

47

was distributed by the American Uniform Boiler Law Society, which stated that the proceedings were based on two theses evolved by the Society: that standardization is the foreword of business efficiency and that civilization and government are based on human life, whose protection is government's first duty.

It was during this period of reaching out and trying to encompass larger segments of business, industry, government, and the public that many specific issues of common import were brought to the fore. Although these were too numerous and in some cases too complex and technical to elaborate on here, a prime example was that of safety valves.

Although crude attempts to devise safety valves went far back in the history of boilers, the concept of "safety" was for so long secondary to efficiency that few inventors or designers cared to spend much of their time on such devices. One of the first documents on the subject in the United States was a government pamphlet, published in 1875 by a Special Committee of the Board of Inspectors of Steam Vessels. This reported tests conducted on valves from selected manufacturers in which "the persons representing these valves were allowed to adjust them to the pressures, supply any deficiency found to exist, and were required to inform the Committee when ready for a test." Unfortunately, in most instances, the valves had to be readjusted after being presented for testing, making it difficult to determine just how effective they were.

In 1898, "Safety Valves," a booklet authored by Richard H. Buel, was published by the Van Nostrand Science Series. This publication not only discussed the concept of safety valves, but devoted many pages and diagrams to present lever lengths and the corresponding weights to meet the set pressure requirements of a given valve. Eleven years later, at an ASME meeting in New York City, "Safety Valve Capacity" was featured, referring to tests to determine lift and capacity and concluding that "nearly all existing rules and formulas are of a kind, which rate all valves of a given nominal size at the same capacity." It was explained that the results indicated that lifts and capacities of valves of different makes, but of the same size and for the same conditions, could vary as much as 300 percent.

The 1914 Edition of the ASME Boiler Code marked the first time in history that all of the makers of safety valves were to agree upon uniform specifications for their products, and the ones that would be most favorable for public safety. However, there was some criticism from the ranks of users and inspectors that the Code was too complicated.

This Edition presented a table for determining the size of valve required, along with an equation in the appendix to use for calculating values not given in the table. These data were based, in part, on the fuel consumed and the heat of combustion of the fuel. The equation also contained a number for the latent heat of evaporations and the coefficient of discharge, but the valve capacity had to be such as to prevent the pressure from rising more than six percent above the maximum allowable. The first edition of the ASME Code also provided design provisions, as well as a maximum blowdown requirement, for which three methods were given to check the capacity: making an accumulation test; measuring the maximum amount of fuel that could be burned and computing the corresponding evaporation capacity; and measuring the feedwater to determine the maximum evaporation capacity.

Because of the industrial developments taking place in America after the turn of the century, the increasing need for technical understanding sparked an escalating interest in higher education. As a result, the number of students enrolled in colleges in the United States in any given year rose from 114,000 to more than double that number by the time the first edition of the Boiler Code was published. Steam generation, necessitating the development of better boilers, was the secret to this new industrialization. Among the milestones of the first decade of the 20th century were the introduction of an advanced boiler designed for pressures of 350 psi; the opening of the first public utility in America to be equipped throughout its system with steam turbines; and the installation of boilers in the power stations of the first New York subway.

It was during this period that the steam turbine earned its place as the major steam-powered source of rotary motion. Among its advantages were its relatively small size compared with the reciprocating engine (described them as "any engine employing the rectilinear motion of one or more pistons in cylinders"); fewer moving parts; low-cost maintenance; and a greater tolerance for overloading. The first public utility plant in the United States to be equipped entirely with steam turbines was the Fisk Street Station of the Commonwealth Edison Company in Chicago. Its eight boilers, each rated at more than 500 horsepower, had gone into operation in 1902. Powering three 5000-kilowatt turbines, they generated enough power to supply what today would be the requirements of about 40,000 homes.

By 1915, boilers for this kind of application had made great strides in design and capabilities. New types of boilers using superheated steam were each capable of generating enough electricity to supply 13,000 modern homes. The race among manufacturers to make improvements during this era was one of the reasons why the Boiler Code evolved at a critical time in America's industrialization. Considerable strides were being made at this time in the design, materials, and manufacture of boiler tubes, which required a relatively thin wall thickness when compared with the "hot-finished" tubes that were readily available. The latter had heavy walls and could be made into light-wall types only by an inefficient amount of rolling or cold drawing. As a result of contemporary developments, hot-finished seamless boiler tubes became commercially available.

Another important development was the introduction of alloy steels, which had been little used by the tubing industry during the more than a quarter of a century in which it had played a part in boiler construction. In the early 1920s, however, the situation changed when several companies, including International Nickel, began experimenting with Monel metal, an alloy consisting mainly of nickel and copper.

During this period, marine boilers were steadily improved, in part because of the escalating need for emergency fleet vessels for service in World War I. As the United States Navy increased its demand for ships, and particularly for its destroyer fleet, the industry saw the first mass production of boilers in its history. Despite such practical applications, critics of the Code asserted that Committee discussions were too theoretical and did not address themselves to utilitarian matters. Such criticism was rarely justified because the subcommittees were regularly involved with specific problems and solutions. For example, during the World War I period, one of the major areas of activity in boiler design was research to solve the mounting problems of caustic embrittlement. Such flaws — it was suspected and later determined — occurred largely because of the higher and higher pressures demanded for boiler operation.

Action in the area of power testing was another relevant example of practical application. In the fall of 1918, the Council, realizing the need for a revision of ASME Power Test Codes, created a standing committee of 25 for this stated purpose. The Main Committee on Power Test Codes with some 20 individual committees was organized in December of that year in cooperation with other societies, including the American Institute of Chemical Engineers, the American Society for Testing and Materials, the Compressed Air Society, and the American Gas Association.

Fred Rollins Low served as chairman. He had started his career in journalism, first as engineering editor of *The Journal of Commerce* and later as the editor of *Power*. He wrote and published books on power, compound engines, condensers, and steam engines at the turn of the century. He invented a number of mechanical devices, including a flue cleaner for vertical boilers, an integrating steam-engine indicator, and a rotary engine.

Another practical development affecting boiler codes was the entry on the scene of the first commercial units known as economizers. These consisted simply of bundles of extra tubes that were positioned in the flue gas stream flowing from the boiler to the stack. As boiler feedwater passed through the tubes, the economizer absorbed heat that would otherwise have been wasted before entering the boiler, and at no additional consumption of energy. The record shows that subcommittees of the Boiler and Pressure Vessel Committee undertook exhaustive studies of economizers and in some cases were able to demonstrate that many claims of economy were considerably exaggerated.

The ASME was by no means alone in its pioneering goals. Indeed, one of the Society's stated aims was to motivate other organizations to examine power problems in innovative ways. The Hartford Steam Boiler Inspection and Insurance Company was one such example, a firm that had worked closely with the Code Committee since its inception. The Hartford pioneered again when, in 1919, it began insuring equipment against breakdowns, including steam engines, internal combustion engines, reciprocating pumps, compressors, and a year later "electrical apparatus of any kind." Each of these new lines of coverage required the services of Hartford inspectors in a concept referred to as "engineering insurance," which signified the trend of establishing closer relationships between technical services and insurance coverage.

World War I brought the United States its first taste of energy conservation, something unique in a society where fuel had long been considered boundless and, in effect, too cheap to bother trying to conserve. This change of pace and outlook in one sense helped to motivate engineers to perfect new forms of machinery in the power industry, where advances in technology were changing the physical operations of the industrial world. One outstanding example was the trend toward larger, more complex plants and, with it, the more extensive use of steam turbine generators and internal combustion engines in place of the familiar steam engine.

This change of direction was accelerated in a unique and unexpected way when the Prohibition Amendment of 1918 clamped the lid on the domestic liquor industry in the United States. The impact in the field of power was considerable since breweries and distilleries had become heavy users of steam. This series of events prompted *The Journal of Commerce* to run an article predicting that the use of steam boilers as a form of power generation would never again rise to its former levels. What the editorial did not foresee was that steam boilers would not only survive but that they would play an even greater role as the primary means of power generation in fossil fuel plants — and later, of course, in nuclear plants. However, the article did pinpoint an existing problem and a trend that was real, if short lived.

Among the major changes in boiler design during the 1920s, as has been mentioned, was the design of larger and larger boilers to increase the capacity of electric generating stations, rather than simply adding new boilers of traditional size. Inherent in this trend was the problem of what to do about existing methods of coal burning, such as stokers, that were becoming obsolete. Insofar as fuel consumption was concerned, one contemporary answer was the development of pulverized-coal firing and the equipment needed for its use. The higher volumetric combustion rates and unit sizes made possible by burning pulverized coal could not have been taken advantage of without the introduction of water-cooled furnaces. They also eliminated internal deterioration caused by slag and reduced the fouling of convection heating surfaces to manageable proportions by lowering the temperature of the gases leaving the furnace.

At first, water-cooling systems were separate entities applied to already-existing boilers. With improvements, however, the boiler surfaces and the furnace water-cooled surfaces were integrated for more effective operation and the resulting unit was referred to as an "integral-furnace boiler." In addition to reducing maintenance, water-cooling also helped users to generate more steam per unit of fuel. Increased steam temperatures and pressures improved the cycle efficiency, reduced the boiler surfaces, and provided for additional super heater surfaces.

As America's demands for electricity mushroomed all across the country, the 1920s saw experiments with the construction of outdoor substations in which the transformers and high-voltage switches were not completely enclosed within any building. By the end of the decade, several steam plants were constructed with the turbines on open decks, but it was not to be until well into the 1930s that utility boilers were installed outdoors, and even in these cases the locations were usually ones in the South and Southwest where the environment was compatible and seldom hostile.

Outdoor installations, as might be expected, challenged Code committees because their rules and specifications had to take into account the lack of protection from the elements and, in quite a few instances, the isolation of locations and the consequent need for remote controls. Yet outdoor installations had certain advantages for public utilities. In the first place, they were much less costly, since the only major structural feature was a concrete slab for the foundation. Equally important, they required far less time for construction and could more easily be scheduled than plants situated inside buildings. In addition, ventilation and cleaning problems were virtually eliminated; stations were considered safer in the event of fire; and all components were readily accessible during accidents or other emergencies.

The major disadvantages were that maintenance costs on boilers, turbines, and other equipment were slightly higher and that extremes of temperature in the area or severe storms could sometimes disrupt operations. Thus, specifications for outdoor components often called for special weatherproofing and waterproofing procedures, especially in the protection of electrical cables, switches, and safety devices.

Just keeping abreast of boiler developments like those mentioned above, let alone formulating rules, was constantly challenging to the Code committees assigned. Yet the work continued during the 1920s with remarkable equanimity and assuredness, covering just about every base imaginable. The mid-1920s saw the evolution of Suggested Rules for the Care of Power Boilers, Section VII, ASME Boiler Construction Code, which had been several years in the making. These rules, used in connection with steam boilers, reported the Committee, will lead to safety in their use. They had been compiled "to assist operators of steam boiler plants in maintaining their plants in as safe condition as possible, the subject of economy receiving only incidental consideration."

The introduction acknowledged "the difficulty in formulating a set of rules that may be applied to all sizes and types of plants" and added that "therefore these rules are suggestive only and it may be advisable to depart from them in certain cases." The five divisions focused on rules for routine operation, operating and maintaining boiler appliances, inspection, prevention of direct causes of boiler failures, and installation.

In all of its publications at that time relating to the construction of boilers, the Committee emphasized that the rules were "not intended to retard the development of new designs or the use of other materials. The Boiler Code Committee will, when so requested, render interpretations on features of construction or other materials as regards their relation to the Code. . . ." As in the past, the Code Committee stated that it "does not pass judgment on the relative merits of particular designs nor does it assume to limit in any way the builder's right to choose any other method of design, or form of construction that conforms to the Code rules."

Among other Code achievements were standard rules for the construction of stationary steam boilers, low-pressure heating boilers, miniature boilers, and boilers for locomotives. Regarding the last-mentioned, a 24-page handbook, *Rules for the Construction of Boilers of Locomotives,* had first been published in 1921, when ASME was still located at 29 West 39th Street in New York City. The work of the Subcommittee on Boilers of Locomotives had been initiated much earlier but had then been interrupted by World War I; and thus a prototype report on locomotives had not been submitted to the Boiler Code Committee until April 1919, at which time it had been discussed at the Spring Meeting in Detroit.

The handbook, 6" x 9" in size, with a blue cover, and listed as Part I — Section III, contained rules for the construction of boilers of locomotives that were not subject to federal inspection and control. It included sections on the selection of materials, strengths of materials, plates, and tubes, maximum allowable boiler pressures, and data about boiler joints, braced and stayed surfaces, and maximum allowable stresses for bolts. The text also discussed safety valves, riveting, fittings, hydrostatic testing, and the use of the official "S" Symbol Stamp and related data.

The spring of 1921 saw the adoption of Rules for the Construction of Boilers of Locomotives, mentioned in the previous chapter. Long delayed because of war interruptions during its consideration and review, it had also run afoul of numerous disagreements among members of the subcommittee. The dissension stemmed from the complexity of specifications for materials, tubes, gages, and various boiler components that had been established by the nation's many railroads. Up until this time, attempts to standardize the working elements of locomotives and rolling stock had been met with frustration on the part of the railroads themselves, most notably those that operated in more than one state and were governed by different regulations, not a few of which were contradictory. Even the American Railway Association and the Interstate Commerce Commission had grappled with the problems and reached only partial solutions over the years.

So the publication of the locomotive boiler code came as a welcome step in the right direction, and at a time when railroads in the United States were headed toward one of the most important eras in their history.

CHAPTER 6

PROGRESS
AND CHALLENGE

Looking Forward and Backward in an Era of Change

Approaching the 1920s, the Boiler Code had proved to be functioning well, although it was still new enough so that the goal of perfection was hardly one that could be realistically attained. One of the disruptive problems that kept arising was the degree of nonconformity among the inspectors in the various Code states. At the meeting of the Boiler Code Committee in September 1919, the subject had been aired and it had been agreed that it would be advisable to form an organization that would be composed of the chief boiler inspectors of all of the states, cities, and regions functioning under the Code.

At this time, a name had been chosen: the National Board of Boiler and Pressure Vessel Inspectors. A headquarters office in New York City had been offered by members of the American Uniform Boiler Law Society. It had become more formalized when, on December 2 of that year, a group of inspectors had met at the office location to form a temporary organization. At this meeting, inspector Joseph F. Scott of New Jersey had been elected the first chairman.

The official formation of the National Board took place in Detroit on February 3, 1921, at a meeting attended by representatives from all areas of the United States affected by boiler rules and regulations. For its part, the Boiler Code Committee agreed that at least one of its appointed members would come from the National Board. C. O. Myers, who was one of the early proponents of an organization for inspectors, became the first inspector to hold that position.

In a paper discussing the beginnings and operations of the National Board of Boiler and Pressure Vessel Inspectors, Walter B. Parker pointed out that the formation of this Board was of substantial significance in the history of the Boiler Code. In its reporting procedures, types of report forms, frequency of inspections, and acceptance procedures, the Board established a uniformity that was remarkable, especially given the fact that, by the time Parker made his report in 1979, it had to cover some 45 states, 32 cities, and five counties, not to mention ten provinces and two territories in Canada.

The National Board, he wrote, "has been so effective that today an inspector can take an examination ... in any jurisdiction that has been approved for giving the examination and, if successful in passing it, can obtain a Certificate of Competency and/or a Commission from the Jurisdiction in which the examination was given, then obtain a National Board Commission." This uniformity of acceptance, he added, "has helped immeasurably by providing qualified inspectors across the United States and virtually eliminated the necessity of inspectors taking multiple examinations."

Frank Williams pointed out that "The ASME was actually hesitant, as a society, about getting involved in all of the complications of inspection and the qualification of the inspectors. As a division of power, it was agreed that the inspection during construction and the final testing would be a responsibility of the states, which ASME agreed to support through its designated group, the National Board. This outlook held true for a long span of years. At the time, it had been evident that there were differences — nuisance differences — between the states and the cities as to their regulations. One of the reasons behind the formation of the National Board had been to minimize this kind of stalemate. The underlying struggle for power was finally resolved through a series of meetings and conferences which resulted in a much better cross flow of communications, more comfortable mutual understandings, and a considerable improvement in the functions of the people and groups involved. This can best be summed up by pointing out that, as we were able to expand the dissemination of first hand information, we minimized the problems in setting rules that would cover the various fields."

It was interesting to consider that, while many of the older laws reflected the thinking that prevailed at the time the laws were passed, quite a few continued to maintain their place in the Code. Testing materials, for example, involved the evaluation of a series of characteristics that would be vital to the performance of those materials in use. Then there was the testing of the *joining* of materials, including those for welding — such as surface tests, X-ray and sonic tests — and in each the objective was to look for internal as well as external characteristics.

"The internal inspection," Parker wrote some 60 years after the formation of the National Board of Boiler and Pressure Vessel Inspectors, "is still the most accepted means of determining the condition of the internal surfaces of vessels; however other means, such as X-ray and ultrasonic examination, are being widely used, particularly where the internal surfaces may not be readily accessible for visual examination. These more modern inspection methods must be recognized in the updated Laws and Rules and Regulations."

Referring to the communications between the Boiler Code committees and the National Board, the foreword states that this kind of cooperation "has been extremely helpful. Its function is clearly recognized and, as a result, inquiries received which bear on the administration or application of the rules are referred directly to the National Board. Such handling of this type of inquiry not only simplifies the work of the Boiler and Pressure Vessel Committee, but action on the problem for the inquirer is thereby expedited. . ."

Another outstanding example of continuing coordination and cooperation relates to the American Society for Testing and Materials (ASTM), which joined forces with the Society in 1919 in the preparation of material specifications adequate for safety in the field of pressure equipment for ferrous and nonferrous materials. Because a subcommittee of ASME had been appointed in 1916 to confer with ASTM and because the two had then cooperated so well, the contents of the 1918 Edition of the Boiler Code had been much more nearly in agreement with ASTM specifications than before. By 1924, the two societies had coordinated their observations to the point where ten of the Code specifications were in complete agreement with those of ASTM, four were in "substantial agreement," and two covered materials for which ASTM had no corresponding specifications.

"It is evident," said the preface, "that many of the material specifications were prepared by the Boiler and Pressure Vessel Code Committees, then subsequently, by cooperative action, modified and identified as ASTM specifications."

One of the fundamentals that characterized the work of these groups in the early days of the Code was the willingness to take risks in order to get objectives accomplished. There is a pertinent chapter about this topic in a current book, *The Social and Cultural Construction of Risk,* published by the Riedel Publishing Company in 1987, most notably in the chapter entitled, "Risk and the American Engineering Profession: The ASME Boiler Code and American Industrial Safety Standards."

Although engineers have always tried to design for safety and the avoidance of danger, their successes have been partial. Nevertheless, in the nearly unanimous judgment of those involved in the technology concerned, the advantages always outweigh the hazards. The automobile is an outstanding example. The challenge, says the author, is posed by technical innovation. Then society is confronted with a new situation with which it must deal. Common sense, rather than scientific expertise, may be the key ingredient. An acceptable response to a technical challenge involves far more than the solution of engineering problems or technical fine-tuning. Social choice is vital.

The book relates that within two years of its founding, ASME had published papers on safety, including the prevention of boiler explosions. In 1883, leaders of the Society argued — without success — for a government program to test materials so that engineers could accurately predict the limitations of equipment they were designing. The initial problem in creating a set of standards was that opposition would come from those who suspected a "conspiracy" by boiler manufacturers against the public — even with the endorsement of a national association.

"The problem then, as now, in producing an enforceable code was that the experts who knew the most about the subject at hand were those who had invented, developed, and promoted the very apparatus that was perceived by the public as requiring legislation." Finding a body that truly represented the public interest was — and always has been — very difficult because of the widely varying interpretation of what exactly is in the "public interest."

The individuals, not the associations, were the ones willing to take the risks — men like Edward D. Meier and John A. Stevens. The Code evolved because some members of ASME "followed a vision of altruistic service to the public good" and because other members were effective in making necessary compromises "palatable to those who must give up something in order to gain a consensus."

As the author said, "The history of the ASME Boiler Code should encourage humility and a healthy uncertainty in those who would solve problems that involve technical expertise but that also involve social and political dimensions." He concluded by stating that the Boiler Code and other such codes "have been a continuing source of pride to engineering societies." However, the solutions to problems of risk and failure will only come about when the public becomes aware and involved. "The public must continue to be the instigators of action, getting first the attention of patrons and then insisting that the engineers assigned to the problems will provide something more than the 'minimum level' of protection from hazards."

Until the early 20th century and the kinds of interest in safety demonstrated in the formation of the Boiler Code, American legislators took the stand that Americans were quite independent and would rather run risks and take gambles than have to put up with government intervention and paternalistic administrations. Ironically, such attitudes derived in part from the fact that professional associations — ASME included — continuously voiced their accumulated opinions that it was better for the private sector to police its activities in a voluntary manner than to have legislation rammed down the throats of the citizens.

Without the enactment of employers' liability laws, it was virtually impossible for employees to collect compensation for injuries sustained in industrial accidents, no matter how negligent the company may have been. Exceptions were rare and if the accident resulted from the ineptness or dangerous habits of a fellow employee, the company was invariably thought to be blameless, no matter what hazardous conditions existed that could have been corrected.

Commercial boilers were often involved in employee casualties and thus bore their share of damnation from the families of the dead or injured. It was acutely distressing that engineers were often in the sensitive position of being "damned if they did and damned if they didn't" when it came to the matter of recommending installations and measures that would protect employees or at least minimize the dangers. As far as management was concerned, the attitude had long been that the job of the engineers was to improve plant efficiency and pare operating costs, not suggest additions that were costly to install and even, in some instances, that required additional power to operate.

It had not been until just prior to the publication of the initial ASME Boiler Code that a concerted effort by the engineering fraternity focused on the need to protect workers in industrial plants. In 1912, a pioneering committee of the Iron and Steel Electrical Engineers managed to obtain the necessary funds and sponsorship for the First Cooperative Safety Congress. The real catalyst was an engineer named Lewis R. Palmer who was said to have travelled more than 4,000 miles just to call on prospective sponsors and solicit attendees who would speak out on the subject.

It was mere coincidence, but the effort paid off the same year that the Code appeared in print with the formation of the National Safety Council in 1915. Its stated aim was to devote itself to "the promotion of safety to human life in the industries of the United States."

As might be expected, the events of World War I dampened many engineering activities that were not directly concerned with military operations, including industrial safety programs. Still, the work of the Code Committee continued within the organization, even if not publicly acknowledged. By the middle of the 1920s, the American Engineering Standards Committee* was able to report that, since 1919, a total of 14 national safety codes had been written and that 26 more were in various stages of completion.

By this time, too, other bodies concerned with standards and specifications had gotten their second wind, to a large measure because the ASME Code had provided the cement that bound members together in common causes and with uniform objectives. In 1924, for instance, the American Society for Testing and Materials, which had been formed in 1898 as "a group of individuals and organizations who feel a common need for better information on, and better means for, evaluation of materials and who recognize that by working together they can more effectively satisfy this need," reached an important milestone. It made its first effort at providing a compendium of information about materials, particularly the fabricating, mechanical, physical, and chemical properties of all of the commercially available metallic alloys required by engineers and other professionals.

ASTM presented answers to the following kinds of questions, among others: How should a product be tested? What tests will the manufacturer apply to know he has produced the specified material? How will the user test it? Is everyone using the same procedure? "There is probably not a single engineering code in the country," said the Society, "that does not make use of some ASTM standards — either materials specifications as in boiler or building codes, or test methods, as in the case of safety codes."

---

*Founded in 1918, the American Engineering Standards Committee became the American Standards Association in 1928, the United States of America Standards Institute in 1966, and the American National Standards Institute in 1969.

Another coordinating body was the American Standards Association (one of whose founders was ASTM), which had been organized in 1918 to provide a nationwide focal point for all standards activities, including materials. As a coordinating agency, its fundamental objective was to avoid duplication in the creation of an integrated body of recognized national standards. It was not structured to prepare standards, but to look to member bodies for the development of standards. Materials standards represented only one of the many fields covered. ASA was the American member body of the International Organization for Standardization (ISO).

During the early and middle 1920s, the industry relied basically on the rules and specifications that had been published in the 1918 Edition of the Boiler Code, even though it had been formulated during difficult times. Many of the materials required for boilers and related components had been classed as critical war priorities and thus were in short supply for commercial civilian use. Despite this, the ASME Committee had steadfastly refused to lower its standards to accommodate substitute metals and materials or to reduce thicknesses and weights in order to conserve supplies. As it turned out, this policy, considered too "adamant" and "inflexible" by some critics, proved to have great foresight. What it meant, in effect, was that the Committee did not have to double its workload by modifying the Code and then having to turn around and rescind the modifications when the war ended. Nevertheless, the war years were foretastes of what would come during World War II on a much larger scale, a period that was partially dormant and when many revisions were either curtailed or postponed.

At the start of the 1920s, the ASME Boiler Committee focused its activities largely on the work of six appointed subcommittees that were active in the areas of locomotive boilers, heating boilers, the qualities of cast iron, rules of Inspection, welding, and the care of equipment in service. There were also two coordinating subcommittees, one of which met occasionally with the Massachusetts Board of Boiler Rules and the other with the American Society for Testing and Materials.

It was the custom of the Main Committee at this time to meet nine times each year, in one- or two-day sessions that were sufficient to cover the agendas at hand. On the average, there would be only five new cases heard, in addition to reviews of old cases being considered for revision or review. Cases were introduced by ASME members or committees as well as by representatives of local jurisdictions, insurers, designers, independent engineers, manufacturers, and users. The procedure closely followed what had been established at the time of publication of the first Code and was found to be both workable and productive.

Among the other ASME rules and specifications that highlighted the period from the end of World War I to the middle of the 1920s were those relating to miniature boilers, defined as shells, no larger than 16" × 42", with 20 square feet of heating surface and 100 psi maximum working pressure; low-pressure heating boilers; steel-plate boilers with pressures of 15 psi for steam and 160 psi for hot water; and cast-iron boilers. Lengthy discussions were held on the various methods of welding then in use and the need for further specifications and on rules for inspection, "which would be suggestive only and not a mandatory part of the Code."

Particular attention was addressed to the subject of unfired pressure vessels. Code Committee secretary C.W. Obert had reported in 1919 that, after an extensive investigative trip, he had discovered that "a great need existed for a code for air tanks and pressure vessels." Some manufacturers had been attempting to apply the power-boiler rules to the design of their vessels, but with only minimal success. Within the next three years, the Code Committee had appointed a subcommittee on the subject, drafted preliminary rules, held two public hearings, and decided that the need for the Code was "urgent."

In January 1922, the Committee voted that "all matters pertaining to the proposed Unfired Vessel Code be turned over to the Subcommittee on Welding, which shall be requested to revise it so as to present a Welding Code to the Boiler Code Committee and to confer as it sees fit with the Executive Committee." It was to be three years, however, before these vessels would receive due recognition, following additional hearings, many more "preliminary" drafts, and continuing suggestions from manufacturers and interested parties of all kinds, including the American Welding Society.

The sixth, and final, draft of the new Code for Unfired Pressure Vessels was adopted by the ASME Council on January 15, 1925, as Section VIII of the Code.

CHAPTER 7

STANDARDS AND STANDARDIZATION
The Evolution of a Science

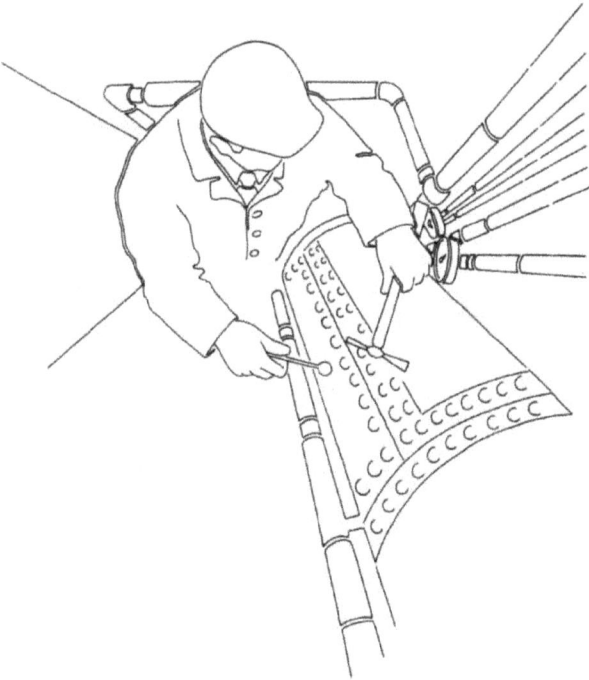

S tandardization," wrote Norman F. Harriman in *Standards and Standardization,* a reference work published by McGraw-Hill in 1928, "in a sense is the bedrock of civilization." Harriman, a member of ASME and the National Bureau of Standards, expressed astonishment that the subject of standardization, in spite of its major importance, had been overlooked by editors and had been largely addressed only in the recorded documents of technical associations.

As he related, the evolution of standards goes back to the crude beginnings of human culture, when community life was governed by customs and common rules administered by the leaders of various groups of people, as well as by the heads of families. Later, the earliest commercial standards relating to professional life were probably those that concerned weights and measures, which were devised and used in Assyria, Chaldea, Babylonia, and — most importantly — Egypt.

The first standards that could in any sense be termed "industrial" were most likely those that were created for the use of raw materials and which ultimately set the pattern for the evolution of the kinds of modern industries we are familiar with today. Specific examples were the spinning of threads and yarns; the weaving of fabrics; knitting; tanning; making footwear, harnesses, and other leather goods; the production of furniture; baking; milling; and preserving foods.

The move to standardization was in part responsible for motivating the invention of four pioneering machines for the textile industry: Hargreaves' *spinning jenny,* Arkwright's *water frame,* Crompton's *mule,* and Cartwright's *power loom.* These machines made it possible for the first time to mass produce goods that were exact look-alikes and promoted the idea of interchangeability.

The invention and development of the steam engine was an enormous milestone in the process of standardization because it greatly accelerated production and made it imperative that parts be interchangeable to prevent slowdowns on the assembly line. By the time his book appeared in print, Harriman was able to state that "Changes in steam power plants are coming rapidly, and the reciprocating steam engine is being replaced by *steam-turbine* units, with higher and higher steam pressures being used. The limit of development of our present stoker-boiler-turbine type of power plant is nearly reached and some revolutionary development is so sure to come that higher and higher allowances for depreciation and obsolescence of equipment should be made.

"The laws of thermodynamics show that more power can be derived from steam if the range of temperature through which it is used is increased. A knowledge of the properties of materials at high temperatures has made possible operations at about 700°F and 3,200 pounds pressure, and such an experimental plant is now in operation in England. At about this pressure and temperature, water reaches a critical point and passes into steam without violent boiling."

Standardization was foreseen in the mid- and late 1920s as the basis of the formula that would help engineers to devise revolutionary new methods for the practical harnessing of power in the future. "One of the most modern aids to industry," said a contemporary report from the American Engineering Standards Committee "is *standardization*, which is now recognized as being of the greatest importance to both producer and consumer. ... In industrial standardization, it is the *consumer* who ultimately is benefitted most, but the immediate benefits are largely to the *producer*, for it at once simplifies his work and enables him to produce what is required by the consumer more cheaply and expeditiously."

During this same period, it was also suggested that mechanical standardization was one of the key factors in the evolution of the United States "from a relatively insignificant position to that of one of the great industrial and producing nations of the world, both in the quantity and diversity of the nation's products."

This long-developing evolution was a cumulative one, as standardization spread from one process to another in the same industry and from one industry to another, until gradually it was recognized that "machines adjusted to standardized work helped one another because of the uniformity of the product."

While recognizing the need, many an engineer liked to quote a saying that was popular in those days and which still carries a strong cautionary message: *"standardization is a useful servant but a bad master."* An introductory comment in a report from the National Bureau of Standards cautioned that Standards should not be allowed to become so crystallized and set that their revision becomes a difficult and time-consuming process.... If improperly used in a rigid uncompromising way, standardization may obstruct progress; if properly used, then standardization will be found to be a reliable and valuable aid to industrial advancement."

In the development of standards — particularly as they affected the Code — it became evident that supportive data were needed on the high-temperature properties of metals used in steam-boiler construction. It was not yet scientifically known how relevant the metallurgical stability of steels and other metals was to maintaining creep strength and preventing the deterioration of mechanical properties through graphitization or other actions. While creep tests had become part of research programs in the 1920s, it was acknowledged that few of these early tests were of long enough duration to provide the desired data.

Conclusions were drawn that high-temperature creep and rupture testing could be "very useful in evaluating the relative strength of steels for high-temperature use, in the development of improved materials, and in permitting the increases in steam pressures and temperatures that have led to greater economy in power production."

It had taken about 50 years for the steam generators or high-capacity boilers of the late 1920s to evolve to their existing state of efficiency and reliability of operation, it was pointed out, and it was a real sin of omission not to know more about the properties and capabilities of the materials whose properties were so critical to the ultimate performance.

"The improvement in boiler performance has come about gradually through the years," said a report by ASME at the beginning of the 1950s, "and is in no small measure due to studies which have been made during the past 25 years relating to the properties and behaviors of metals under continuing exposure to stress at elevated temperatures."

So even as late as the mid-1930s, this kind of long-range testing was still in its infancy. Yet it was even then recognized that such research and investigation would pay off in terms of superior materials of construction for the superheater of the future, where creep became operative.

Because of the increasing significance of pressure vessels and their components in the evolution of the Code, long-term tests were also recommended for these products and materials. By the middle of the 1930s, progress in pressure vessel design and fabrication since the publication of the PV Code a decade earlier had brought about marked changes in the form of bolted connections. Whereas flanges had formerly been provided with hubs of approximately uniform thickness, screwed to or slipped over the shell or pipe, it had by then become customary to use tapered hubs and obtain integral structures by butt welding the end of the hubs to the shells. Thus, the traditional methods were becoming obsolete.

During this decade, methods of construction also underwent radical changes. Earlier flange designs had nearly all been based on the principle of incorporating hubs of uniform thickness. But the widespread acceptance of welding had brought about the changes, most notably the tapered hub mentioned above.

Methods for calculating stresses in these flanges, however, had been found to be very limited, varying with the unit, the size, and other specifications. The industry was still relying heavily on what were known as the Taylor-Waters formulas for determining stress, which had been presented by ASME in 1927 and which were considered "reasonably accurate." Following that step, flange design had become a subject of intensive discussion, somewhat muddied by the very numbers of solutions presented by would-be contributors. The approach taken by ASME was quite traditional: the appointment of a subcommittee to investigate the subject. Thus it was that representatives were selected in 1934 from the Boiler Code Committee, the Joint API-ASME Pressure Vessel Committee, and the American Standards Association. They were charged with the task of drafting a set of rules for uniform procedures that would cover the "all important phases" of the design. This action was described later as "the first known effort to introduce a rational stress analysis into the commercial design of flanged connections."

Developments like these might have appeared on the surface to be highly specialized and little related to products and materials in completely different categories. But what engineers and researchers were finding out every time they conducted tests was that many parallels and overlapping principles existed that ran right across the board. Equally important, there were connections — sometimes not evident — between many codes for unlike products and materials.

One of the most productive means of conducting research with this kind of overview and with strong objectivity had been found to be through collaboration with the growing body of independent, non-profit organizations interested in technical research. One of these was the Engineering Foundation, self-described as "an institution of engineering and scientific research whose goals are the furthering of research in science and engineering or the advancement in any other manner of the profession of engineering and the good of mankind."

The Foundation had been established in 1914 through the efforts and considerable financial contributions of Ambrose Swasey, who had enjoyed a long and successful career as an engineer in New England, Illinois, Ohio, and New York. A past president and one of the founders of ASME, he had long been concerned about the need for research in many fields of engineering. He had also been instrumental in the founding of the National Research Council in 1916, which performed notable service during World War I. By the mid-1920s, the Foundation had become involved with a number of major projects that were relevant to the work of the Code committees, including metal fatigue, internal stresses in metals, residential heating boilers and fuels, the generation and distribution of power, the problems of increasingly high-steam pressures, the strength of metals at high temperatures, the corrosion and erosion of metals subjected to high temperatures, the stress of metals in boilers and pressure vessels, the graphitic corrosion of cast iron, and welding methods and procedures.

In all of these areas of research and testing, it was constantly emphasized that the finest products and materials that could be produced were useless in the long run without constant and reliable procedures for *inspection*. The 1920s thus became a proving period for the industry, with the active support of insurers, the National Board, and all who had a direct stake in the improvement of inspection techniques. There were many advances, but also a few setbacks in the form of serious and sometimes catastrophic explosions. It is not by chance that the history of boiler inspection was colored by references to accidents. More important than the events per se was the fact that such disasters often had a positive effect in that they motivated engineers, representatives of industry, and government to establish codes that would reduce the number of casualties and minimize property damage.

From the very beginning, visual testing was an important part of boiler inspection, since well-trained inspectors could in many instances forestall the devastation of steam before accidents occurred. By the middle of the 1920s, nondestructive examination was becoming more sophisticated and more effective. By then, too, the mechanics of steam's vast destructive forces were more fully and widely understood. It had been determined that water heated to the danger level and suddenly uncontrolled would flash instantly into steam with an extremely large volume increase — as much as 1,700 to one.

"A faulty boiler is a time bomb," explained a recent editorial in *Materials Evaluation*. "A sealed container with 475 L of water at 149°C holds as much energy as 2.3 kg of nitroglycerine. It is no wonder that, in the 19th century, the devastation caused by boiler explosions was widely thought to be an act of God. . . ."

After the formation of the National Board of Boiler and Pressure Vessel Inspectors in 1919 by a group of concerned state enforcement officers and others, continuing research and study brought to light more effective ways of testing boilers in a nondestructive manner. At first, many of the methods were still somewhat primitive or personalized. One inspector, for example, described his "finger test," used to judge the thickness and corrosion of steel plate. On one occasion, after using this test, he informed one user that a boiler was defective. Ignoring the warning, the operator fired up the boiler, which exploded two days later, causing much damage and killing a nearby horse.

Later in the 1920s when manufacturers began using fusion welding quite commonly instead of riveting in the fabrication of boilers, another form of nondestructive testing was applied known as "tapping." Inspectors tested the welded seams by tapping them with hammers and at the same time listening to the sound through a stethoscope — much like the procedure used by physicians on the human body. A sound that was "dead" indicated a defective weld.

During this period, magnetic particle testing and dc induction testing were tried out but evidently with mixed results.

Radiographic testing (RT), however, which was experimented with in the mid- to late 1920s, looked promising. An ASME subcommittee was selected to define the standards and by 1931, the revised Boiler Code accepted welded vessels that had been judged safe by RT. Subsequently, boiler manufacturers began to build vessels that were much larger and that could be subjected to greatly elevated temperatures. By this time, magnetic particle testing (MT) had finally come into its own because of its capacity for detecting surface cracks that were missed by radiographic testing.

Inspection and testing had come a long way, but they were still rudimentary in many respects.

CHAPTER 8

THE DECADE OF POWER
The Responsibilities of World Leadership

Too often the 1920s are associated with frivolous subjects like bathtub gin, the raccoon coat, and rumble seats. In fields of technology, however, remarkable developments were taking place that were often overlooked by the public. One promising example was the ready availability of steam power, a major factor in spark-plugging an unprecedented national prosperity. America was now using almost as much electricity as the rest of the world's nations combined. Of the 27 million autos in the world, 22 million were to be found in the United States, not to mention 61 percent of all the telephones and a plurality of the major products and machines using electricity and other forms of energy.

Among the most significant happenings were the design and production of boilers and related equipment and accessories, along with the improvement of metals that could withstand the increasing temperatures and pressures. Steam generation reached new plateaus as the ever-expanding demands of home, office, and assembly line for additional electrical power resulted in rapid improvements in central station design, especially to accommodate the increasing use of higher steam pressures and temperatures. During the 1920s, steam pressures rose from less than 400 pounds to more than 1,400; steam temperatures jumped from 650°F to 750°F; and the number of pounds of coal required to produce a kilowatt of electricity fell from just under three to about 1.6 because of the improved efficiency of boilers used by public utilities.

These major changes in boiler design and construction were seen most notably in the trend toward larger boilers rather than, as had been the practice, the addition of new units to meet increasing needs. As a result, engineers also had to devise new methods of firing them when older equipment, such as stokers, became obsolete. One solution was to pulverize coal by grinding it to a fine powder. When mixed with air and blown into water-cooled furnaces through openings in the walls, this form of coal proved to be not only easier to handle but considerably more productive in delivering heat.

Advances like these had to be continually monitored by the ASME Boiler and Pressure Vessel Code Committee because they necessitated new approaches to the design and construction of furnaces and related equipment. Research also had to be underwritten and tests conducted to determine whether pulverized coal, which resulted in a form of fused ash not common to lump coal, caused corrosion, erosion, or other deterioration in metals.

In the new types of water-cooled furnaces, some of the tubes forming the circulation system of the boiler extended downward along the walls of the furnace. That placed them in the active combustion zone, where the water absorbed the intense heat that resulted from the burning of pulverized coal. Water circulation was designed to cool the walls, while at the same time the absorbed heat generated steam. By cooling the gases before they entered the boiler bank or superheater, the water-cooled walls also reduced other operating problems.

The Code Committee had to review the standards and rules for tubes, among other products, as a result of this kind of changing power technology.

Instrumentation also became more important during the 1920s to meet the needs of increased power generation. At the time the first Boiler Code was published, power plant operators required very few instruments other than simple water-level and pressure gages. However, as temperatures and steam pressures increased, and as boiler sizes expanded, instruments for measuring performance and other characteristics became indispensable.

"The Code Committee did not initiate the kind of research and development we are talking about here," explained Frank Williams, recounting his lifetime career with Code work and related functions. "But the fact that the Committee was, by nature of its procedures and fundamental thinking, able to deal with all these matters as they were introduced is the real story. Historically, we see an interesting phenomenon occurring over the years: the creation of organizations that were motivated by the Code Committee and its needs, which addressed themselves to new fields of research."

Developments in the design and performance of steam generating equipment had reached this plateau only after a half century of pioneering, experimentation, and improvement. The evolution during the 50 years from 1880 to 1930, for example, was divided into three overlapping periods and methods: hand-fired, stoker-fired, and pulverized-coal-fired. In the latter part of the 19th century, the cost of fuel had been so low that investment in more efficient equipment received scant consideration. By the 1920s, however, the picture had changed drastically and, largely because of the high percentage of steam plants used to generate electricity on a competitive basis, the investment in more efficient and higher pressure equipment became all-important.

The introduction of the steam turbine came at the turn of the century, followed by its rapid development as applied to public utility plants. Contemporaneous with this came the enlargement of the boiler and a greater rate of fuel input per width of furnace. In 1923, four boilers were installed in the 59th Street Station of the Interborough Rapid Transit Company in New York City. Though occupying the same floor space and having the same furnace width as those installed in 1902, they had a capacity of more than four times that of the older units.

Pulverized coal was experimented with as early as the middle of the 19th century, but no real success had been made prior to 1916, when the first practical installation was made for the Missouri, Kansas, and Texas Railway Company at Parsons, Kansas, as fuel for eight 250-horse-power boilers. Four years later, pulverized-coal-burning equipment was installed at the Lakeside

Station of the Milwaukee Electric Railway in connection with eight 1,306-horsepower water-tube boilers. This was recognized as the first major central station installation of pulverized coal in the world situation for the general fabricator.

Many fabricators did not fully understand the Code or the reasons behind its existence, said Robert Chuse, author of several reference works on unfired pressure vessels, adding, "There is no substitute for reading and understanding the ASME Code. Many protests are heard on this score, even from inspectors and engineers, but a knowledge of those sections of the Code that pertain to the type of work in which one is engaged is a necessity." As the author pointed out, some vessel requirements were extremely simple, while others reflected the "vast industrial complex" that had created a wide variety of unfired pressure vessels which constituted "a perpetual challenge to the ingenuity of the fabricator. . . . Since its first edition in 1925, the ASME Unfired Pressure Vessel Code has been an important reference for designers and fabricators of pressure vessels. Its specifications now govern all unfired pressure vessels used in most states and all Canadian provinces."

According to an article in *Materials Evaluation,* May 1986, pressure vessels as a class suffered from numerous flaws until the early 1920s because of poor design and fabrication and a lack of adequate rules and specifications. "At the close of World War I," explained the editorial, "there was a rapid advance in methods of construction, especially of pressure vessels. The most important development was the introduction of welding and its extensive use as a substitute for riveting."

Paralleling the milestones in the design and production of steam generators and boilers were the advances being made in the field of pressure vessels, culminating in the Code for Unfired Pressure Vessels that was published in 1924. For many years, pressure vessels had been something of an orphan in the industry as far as codes and standards were concerned, and there had been little concerted effort on the part of designers and manufacturers to resolve the situation. The producer of pressure vessels was faced with serious problems, explained Robert Chuse in a book published in 1954, *The Unfired Pressure Vessel Code Simplified.* Severe competition existed throughout the fabricating industry, "not only because of the growth in the number of fabricating shops but also because of the substantial technological advances made in recent years. Many of the older, more heavily capitalized plants have been hard hit. Some of these have successfully turned to specialized fields; others produce specialty products that return a good profit when in demand. When demand slackens, these shops often turn to general fabrication, thereby increasing the competitive rapid advance in methods of construction, especially of pressure vessels. The most important development was the introduction of welding and its extensive use as a substitute for riveting."

Although early welded vessels were frequently unsound, improvements in the field of welding were very rapid during this period, especially at the time metal arc-welded boiler drums were introduced around 1925. This era saw not only such technological advances but also significant developments in engineering insurance that were later to influence Code inspection procedures. As noted earlier, insurance programs had been created much earlier in England than in the United States, culminating in 1917 in the formation of the Associated Offices Technical Committee (AOTC), whose members consisted of the top boiler insurance firms in Britain. Upon the formation of the British Standards Institution, AOTC had been represented on the committees charged with the preparation of standards for various types of pressure plants.

In the United States, however, the insurance business was undergoing many upheavals and facing new challenges, not the least of which was trying to cope with the changing nature of the boiler industry as it started producing fewer boilers but ones of larger size and with greater capacities. Other factors that impinged on the nature and extent of the boiler insurance business during this decade were the increased use of internal combustion engines in smaller industrial operations, the advent of the large central station for the production of electrical power, the gradual acceptance of oil as a fuel, and the increased use of the automatic stoker.

The boiler insurance business was greatly enhanced, however, by the ever-widening acceptance of boiler codes and standards and the consequent reduction in accidents, casualties, and property damage. As a result, by the end of the 1920s, there were sharp reductions in premium rates on boilers and pressure vessels, ranging from 15 percent to as much as 35 percent. This era also saw the introduction of new forms of commercial and industrial insurance. Among these were electrical insurance, to take into account the increased generation of electricity; furnace explosion insurance, to cover new problems arising from the presences of gases in furnaces and the flues of power and heating boilers, and power interruption insurance, to provide indemnity for any loss of power supplied by a public utility.

During the decade following World War I, there were many other events and highlights of significance to the evolution and use of the ASME Boiler and Pressure Vessel Code. Among them were the following.

•Cooperation and Coordination With Canada

Cooperation with Canada started at an early date in the history of the Boiler Code, with origins in the 1920s, which eventually resulted in a published "Agreement of Cooperation Between the Engineering Institute of Canada and the American Society of Mechanical Engineers." Key points were that the

International Council would be comprised of equal numbers of representatives from the ASME and the EIC and would meet regularly, at least on an annual basis; that joint meetings would be held on specific matters of mutual interest; and that student organizations would be encouraged to become actively involved in ASME Student Branches and EIC Student Sections.

The International Council was charged with reviewing the plans and programs of each organization and making recommendations for broadening the activities of one so they may be of greater value to the other; close participation by members of one in those activities of the other which may be useful to either or both, and joint projects that may be mutual to the engineers of Canada and the United States.

•High-Temperature Developments

Through the use of alloy steels in superheaters in the late 1920s, steam temperatures above 750°F — until then considered "the maximum temperature practical" — became possible. By 1930, manufacturers were accepting contracts for temperatures up to 900°F, with indications that temperatures in the 1,000°F range were practical. The earlier-mentioned trend toward larger and larger water-cooled furnaces and steam generating units meant, among other things, that it was by then possible to construct single units which could furnish the steam required for 100,000-kilowatt turbines.

•Power TestCodes

During 1921, the Power Test Codes Individual Committee on General Instructions and Reciprocating Steam Engines completed the revisions of these test codes, after several stages of development. The Committee, which consisted of 125 specialists, had been engaged in revising the codes of 1915 for the previous three years. A public hearing had been held and most of the rules had then been adopted "with such slight verbal changes as the Committee might find were necessary for the sake of consistency."

•Technical Committee Reports

During the middle of the decade, it became the custom to summarize the codes and standards completed during the course of each year, make them available in pamphlet or book form, and publish the preliminary drafts in current issues of *Mechanical Engineering*. At this time, they included reports on such topics as the inspection of materials and boilers; the Code for unfired pressure vessels; rules for power boilers; material specifications; test codes for reciprocating steam engines, internal-combustion engines, and steam turbines; and various interpretations of the Boiler Code.

The Reports also included a section on research, which covered, among others, publications on fluid meters; properties of steam; riveted joints; boiler feedwater; and boiler-furnace refractories.

A number of changes in personnel occurred during the 1920s that were meaningful in the history of the Code. John A. Stevens presented a letter of resignation, calling attention to the fact that he had served on the Committee since 1911, during the crucial years of the Code's creation. He was persuaded to remain one more year when so petitioned by a resolution of the members and finally resigned in September 1925. Fred R. Low was appointed chairman to succeed him. Two years later, C.W. Obert also resigned, having served as the Committee's secretary for 16 years. He was appointed Honorary Secretary and a member of the Boiler and Pressure Vessel Committee, to which he had never officially belonged.

CHAPTER 9

PUSHING THE LIMITS
Increases in Pressures and Temperatures

By the beginning of the 1930s, following the stock market crash and the radical changes taking place in the world of investment and finances, the ASME Council, the Code Committee, and all administrators involved in the Society's codes, standards, and certification, had to sit back and review their own policies in a society shaken by these upheavals.

It had long been accepted, almost without a second thought, that the work of technical divisions could not long continue, or even survive, without strong industry support. This fact was markedly evident in the planning, creation, and publication of the Boiler and Pressure Vessel Code. How could committees be formed and assured the necessary blocks of time in which to function without the assistance of organizations within the industry? How could volunteers be granted freedom to take leave of their jobs for substantial periods of time in order to serve on subcommittees? How could specialists, often in short supply, be released by private industry so that codes in the making would benefit from the advantages of talent and experience?

Although some members of the Committee at this time, notably Dr. David S. Jacobus, opposed the idea of appealing to corporations for monetary support, it was no secret that several large engineering firms, such as Babcock & Wilcox and Combustion Engineering "indirectly provided incalculable amounts of financial assistance." Such firms recognized the fact that there was much to be benefitted from cooperation and mutual assistance, far beyond any financial recompense. Boiler and pressure vessel codes provided many advantages to industry, over and beyond the factor of uniformity. There were cases, for example, in which changes in a particular code had helped to solve certain industrial manpower shortages without recourse to higher wages.

That kind of strategy sometimes helped also to avoid conflicts and jurisdictional disagreements between the various crafts engaged in the design, production, and installation of boilers and related equipment. The creation of codes and the consequent organization of groups like the National Board of Boiler and Pressure Vessel Inspectors produced what was at one time described as a practical blend of both idealism and self-interest. This effective balance assured boiler manufacturers and other companies in the industry that they could remain solvent and profitable, while at the same time recognizing and acting upon public concerns for safety and dependability.

The procedures were initiated by private industry under the assumption that those in any given industry best knew the economic and technological requirements and were in a position to play a major role in defining codes and standards. It was vital that industry take this stand because it was evident that government in the United States was going to take few steps to create standards unless forced into it by the private sector. The checks and balances were assured by the fact that the representative interests in achieving these codes and standards included consumers and the general public as well as the producers and engineers who had more of a vested interest in the outcome.

Neutrality was a motivating factor in the overall intent to create codes and standards that were to the best interest of all concerned, supported by effective measures and a sense of professional purpose. The result was that the Boiler Code achieved the right kind of recognition during these years and would be adopted as law in most states and Canadian provinces. Described as "a model voluntary standard," the Code was "often used as an example of the ability of the private sector to generate standards that served the public interest."

In the period from 1931 until 1934 Dr. Jacobus promoted programs by the Main Committee and all other Code committees to encourage members to communicate more clearly with each other. This represented an expansion of functions and a considerable change in outlooks. In the past, many subcommittees and their members had tended to take a defensive, if not hostile, attitude. This was quite natural, since there were often points of controversy in the discussions about revisions and the inclusion of specific words and phrases. This new spirit of cooperation and coordination was fostered by another, basically social, development. Because of the success of an experimental meeting held at a country club in New Jersey, an attempt was made to hold further gatherings in atmospheres that were more relaxed and conducive to congenial discussion than the usual committee halls. During this period, tours were organized so that members could visit with each other, not only in New York City, but in Houston, Los Angeles, and cities across America. Tensions were eased and communications were markedly improved.

Along the historical road from past to present, the Boiler and Pressure Vessel Committee has traditionally displayed one workable philosophy that has helped to assure the success and steady growth of the Code. That has been its affinity and close relationship with other non-profit associations with similar objectives. One of the best examples is that of the ASME-ASTM Joint Committee on Effect of Temperature on the Properties of Metals. The Committee was founded in 1925, following a symposium on the subject that had been held the preceding May. The subject was prompted in part because of the continuing rises in the operating temperatures of steam turbines and a number of technical papers pointing out the great savings possible in heat rates and coal consumption as a result. Thus, it was a timely subject that merited discussion and review.

By that time, an ASME research program in this field was already underway. Although most tests were for short-time exploration, lengthy tests of several thousand hours duration were cited as being feasible and with a greater probability of accuracy. While creep at high temperatures and failure after a period of time, for example, were known to occur, reliable data were very limited. Later, progress was made in a number of other technical areas, including comparative high-temperature tension tests on carbon steel, the matter of thermal expansion (1927), a machine for investigating creep tests at high temperatures (1929), and the correlation of technical data (1930). Some tests were of long duration, two months or more.

The committee was enlarged in 1930, reorganized into four subcommittees: organization and membership, finances, technical projects, and technical data. In June 1931, there followed a joint meeting of ASME and ASTM in Chicago, the first symposium organized by the Joint Committee, "Symposium on the Effect of Temperature on the Properties of Metals." Among the subjects covered were: engineering trends and requirements for metals at high and low temperatures; requirements for boilers, turbines, and steam piping in the oil, chemical, automotive, and ceramic industries; the properties of metals available for high- and low-temperature service; thermal expansion and conductivity; and the properties of rare metals.

At that time, a dozen or more projects were in the works, including material procurement; short-time tensile tests; impact tests; fatigue tests; elastic measurement and shear; wear and seizure; chemical stability; and structural stability.

In 1933 a new committee was formed, known as Committee V, to study oil refinery problems. In addition to members of the joint committee, this had representatives from the American Petroleum Institute (API) and the American Association of Steel Manufacturers (AASM). Reports were made on such subjects as tensile strength of metals, yield strength, elongation, and breaking strengths under pressure and high temperature. Even though a number of findings were inconclusive, they were quite innovative for that era.

"Most of the problems in the oil industry were considerably different from those in the boiler industry or those related to air tanks and smaller, lower-pressure vessels," recalled Frank Williams, who was involved in petroleum matters at this time. "When the decision was made to develop the joint API/ASME Code, the engineers were working in both camps because designers and manufacturers were concerned with both industries. There were several matters of importance. One was the stress factor. Should it be a factor of four? The oil people thought it should be three. Another question was how high the test pressure should be. When we finally came to grips with the subject, it was decided to form a Committee on Code Stress Basis."

The year 1933, reported "A 40-Year History of the ASME-ASTM Joint Committee on Effect of Temperature on the Properties of Metals," was also noteworthy for the first publication of the tentative codes for Short-Term High-Temperature Tension Tests of Metallic Materials and for Long-Time (Creep) High-Temperature Tension Tests of Metallic Materials.

Extensive changes were also made in codes relating to high temperature and tension creep tests. There was continuing preoccupation in those years with codification, one example being that no less than ten laboratories were commissioned to run high-temperature tension tests on a particular lot of low carbon steel to "establish the adequacy of this test method." Seven additional laboratories were involved with similar tests to check out the creep code. "In retrospect," says the record, "one asks if the 17 laboratories could not more profitably have devoted this effort to testing new and better materials."

The second decade of this ASME-ASTM Joint Committee was referred to later as "the creep-test years," touched off by the announcement that prolonged creep tests of the order of 25,000 hours had been scheduled under the sponsorship of the Joint Committee at Battelle Memorial Institute. In 1936, a report was made on creep tests that exceeded one year's duration on steel containing various proportions of chromium and nickel. By this time, distinctions were being made in the method of reporting on the tests. It was pointed out, for example, that very careful use of language had to be considered because there was a difference between "a creep rate of one percent per 100,000 hours" and "a total creep of one percent in 100,000 hours."

It was recognized, as a result of such tests, that "metals might have a life that comes to an end after a sojourn at stress and temperature."

Speaking about the composition of the Joint Committee, member Ernest Robinson stated that "there is no doubt that here were groups of expert and dedicated men, the most competent that could be gathered together, working on a mission of great importance to American industry and, after the War came, of great importance to the military might of the country." He also emphasized that the work of the Joint Committee "has had great educational value for all its members, whether learned scientists or get-it-done engineers, by bringing together important men of widely different points of view."

1938 "was a milestone year in the history of the Joint Committee." The Committee had grown in stature over the years and had begun to command respect and wide support from industry. Accurate and reliable test methods had been perfected and were well recognized, as was the scientific analysis being directed to all the activities of the organization.

That year, the creep code, first released in 1933, was published with extensive changes and was "at last being forged into a shape that was to endure for a long time." At this time, tubular creep tests and torsion creep tests had been started at the University of Michigan and were to be successful, yielding results for the analysis of strains and stresses under complex loadings.

Furthermore, a project on manufacturing variables was also to begin its stated objective of eventually convincing the manufacturers of materials for high-temperature service that every detail in the manufacturing process was vital. This would dispel the long-accepted myth held by both manufacturers and users, said Robinson, that "a simple statement of chemistry and heat treatment should suffice to distinguish between the better and the poorer."

"Perhaps the most finished achievement of 1938 was the publication by ASME and ASTM under the sponsorship of the Joint Committee of the great volume entitled Creep Data . . . ." This was 848 pages long, complete with logs, charts, and diagrams, and supporting data that even included photo micrographs. There were 24 sections on wrought steels and ferrous alloys, nine sections on cast steels and ferrous alloys, and four sections on nonferrous alloys.

With the onset of World War II, the Joint Committee continued its research and testing, not only with universities and the Bureau of Standards but now on many occasions with the War Department and others directly concerned with materials and equipment that would be used in the war effort.

It was inevitable that the committee would have to address itself to the subject of graphitization since, "war or no war, the power stations of the country had to be kept running." Carbon steel pipe at temperatures over 800°F had shown a disturbing tendency to graphitization in the region of welds. The replacement and rewelding of steam piping in the power industry had been the main solution to the continuing problems and in 1943 the Joint Committee sponsored a session at the ASME Annual Meeting on the subject, followed by programs of research and testing.

"The material that was suffering from this disease," said one report, "had been placed in service years before with every expectation that it would have satisfactory high temperature strength, but then it began to develop isolated spots of weakness and cracked in service."

At the beginning of the 1950s, a new subcommittee was organized, this time to study problems at extremely low temperatures, especially in petroleum and chemical operations. Studies of the brittle behavior of metals also included those subjected to low temperatures. At the other end of the spectrum, another group was involved with a study of the corrosion of materials when subjected to elevated temperatures for varying lengths of time. Strength and ductility at high temperatures was also tested. In 1953, ten years after it was begun, the study of graphitization reached some conclusive — though not final — findings. It was reported, for example, that aluminum promoted graphitization while nitrogen increased resistance to it, and that 0.25 percent chromium steel graphitized more readily than carbon steel.

A noteworthy trend in the mid-1950s was the Committee's collaboration on international research projects. There were two programs, for example, which studied the comparative high temperature properties of British and American steels, and the examination and testing methods used in Great Britain and the

United States to determine creep stresses. This was followed in 1959 by recommendations to sponsor an international conference on creep and rupture, which then came to fruition in 1963 as an International Symposium held in New York and London.

Focusing on the growing importance of materials used for missiles, rocket engines, and supersonic aircraft during this era, studies were also conducted to provide data about rupture and deformation of various alloys. These ranged from changes that occurred in less than a minute to those of 1,000 hours duration at temperatures from 1,350°F to 1,800°F.

During the first 30 years of its existence, the Joint Committee had sponsored 16 Special Technical Publications, most of which were published by ASTM. During the next ten years, however, 16 more STPs were published, indicating not only the escalating activity of the Committee, but the increased interest of private companies, which participated in programs and supplied financial grants, as well as donating materials and manpower.

On October 4, 1957, the Space Age began with the abrupt announcement that *Sputnik I* was in orbit. Less than seven years later, the whole basis of the science of navigation and of worldwide communication had been revolutionized. At the meeting of the Committee in 1960, a paper was presented by Harold Hessing of NASA that outlined the problems in the design of a jet-propelled transport plane intended to fly at three times the speed of sound. He solicited the help of the Joint Committee. It was a unique milestone, for no government agency had ever directly sought the help of the Joint Committee before — a sure indication that this group had attained a greater stature and, indeed, that some government agencies had attained greater sagacity!

Flashing back to the 1930s when this highly productive and innovative joint effort was born, we can trace similar — if perhaps less dramatic — examples of reaching conclusions to difficult problems through mutual effort and cooperation. This was a period when individual and group participation in matters relating to the Code was burgeoning. By the start of the 1930s, the Code had grown to eight sections:

I   Power Boilers
II   Material Specifications
III   Locomotive Boilers
IV   Low-Pressure Heating Boilers
V   Miniature Boilers
VI   Rules for Inspection
VII   Suggested Rules for Care of Power Boilers
VIII Unfired Pressure Vessels

The introduction had remained largely unchanged, brief and emphasizing the Committee's policy not to take any steps or publish any rules that would retard the development of new designs or the use of other materials. It was also reiterated that "the Committee does not pass judgment on the relative merits of particular designs or construction."

By this time, there were no less than 20 different power test code subcommittees and almost two dozen groups dealing with safety in one aspect or another. All in all, the ASME and related groups could point to a total of more than 200 committees that were working on the subject of standards in the mechanical engineering field alone. The specifications for materials in the Code were constantly the same or similar to those of the American Society for Testing and Materials, with a very few exceptions, mainly products or materials for which ASTM had no corresponding specifications.

Both ASME and ASTM constantly monitored advances in the field, as well as in the laboratory. The continuing quest by public utilities for greater efficiency during the 1930s was reviewed in great detail every time a pioneering effort, no matter how limited, proved to be successful. Modifications of the conventional steam cycle were tried again and again. One of these, which actually was based on a binary fluid mercury steam unit operating at a New England utility, used high-temperature low-pressure mercury vapor to top a conventional steam cycle. Binary fluid topping cycles were so named because the rejected heat from one fluid cycle was used to supply heat to another fluid operating in a lower temperature range. (Hence the use of "binary," to indicate a paired operation). In the mercury-steam cycle, the mercury condenser also acted as the steam boiler. Other high-efficiency cycles that were tested involved combinations of gas turbines and steam power and direct thermal-to-electrical energy conversion. Yet, in spite of the many complex cycles devised to increase overall plant efficiency in the 1930s and later, in the end the conventional steam cycles proved to be the most economical, at least right up to the mid-1970s.

Regardless of the success or failure of attempts like these to improve efficiency and economy of operation, ASME and the other organizations it coordinated its activities with had to do their homework to determine what factors might, or might not, be relevant to the Code. The same would be true, in a far more exhaustive and complex way, when atomic energy appeared on the horizon. And, as early as the end of the 1930s, a few members of the Society were beginning to give thought to this revolutionary phenomenon, even before that historic date in 1942 when Enrico Fermi created the first self-sustaining chain reaction in uranium in Chicago.

Design considerations were not studied until later, of course, when the Boiler Code first undertook the challenge of adding nuclear specifications in a manner similar to those for fossil-fuel boiler units. However, it had been recognized much earlier that special emphasis would have to be focused on safety in light of the very nature of nuclear energy and the fact that severe accidents could cause many more casualties and greater damage than would result from accidents in plants using conventional fuels.

By the beginning of the 1940s, certain basic facts about the future of atomic energy were already becoming known. For one thing, it was understood that, because of the great differences between nuclear systems and those using fossil fuels, there would eventually have to be wide variances in the makeup of the codes and rules for power generating equipment. It was clarified that in nuclear systems the temperature difference that would produce the flow of heat between the primary and secondary water would be dependent upon the difference in pressure of the two fluids and the hotter primary fluid would be maintained at the higher pressure. Therefore, a steam-generator design using a minimum amount of heat-absorbing surface would have to have the boiling secondary fluid on the outside of the tubes, since a tube, or cylinder, could withstand greater internal pressures. This contrasted with water-tube boiler designs that are fired by fossil fuels in which the boiling liquid is contained within the tubes.

It was also estimated that heat-transfer coefficients would be high in nuclear steam generators and consequently, in most cases, the tube metal would be important in controlling the overall rate of heat transfer. Thus, tube diameters would have to be small and tube walls as thin as practical.

CHAPTER 10

THE "GREAT PERIOD"

Overcoming the Ravages of the Depression

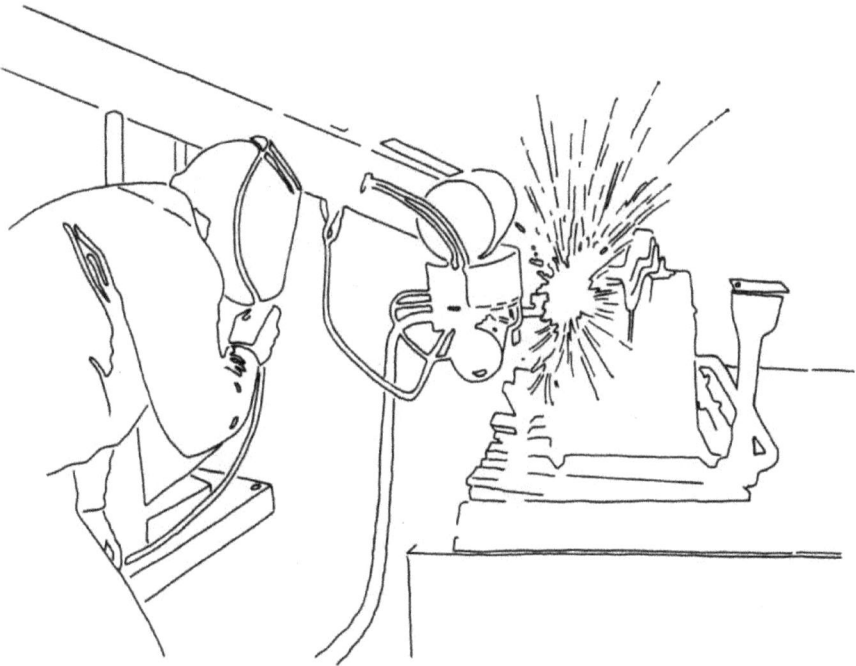

T he national economic crisis that resulted from the Crash of 1929 actually had a unique "annealing" effect on the industry and motivated diversification and new engineering developments. "On the industrial scene," reported one economic commentator, "the Depression was an ill wind that blew frequent good." For one thing, the financial problems faced by most companies and industries challenged them to tighten their belts, strive for greater efficiency, and try to attain a better yield for each dollar of investment.

This accelerated search for efficiency and economical energy paid off in the power field with some developments that had far-reaching results. Significant advances were recorded, including improvements in steam generating equipment for both land and sea; new designs that made possible the increase of pressures from 1,400 pounds to as high as 2,500 pounds and increased temperatures up to 960°F; metallurgical research in the field of tubular products that constantly pushed back heat-and-pressure design limitations; and new types of firebrick with ultra-high insulating qualities which virtually revolutionized industrial furnace design and construction.

One of the major breakthroughs in boilers in the 1930s was the fusion-welded drum. Prior to 1930, the standard method of joining boiler drum plates was by riveting. The butt straps and extra strengthening collars that were required in riveted construction not only added weight to the drum, but demanded constant maintenance and diligent periodic inspection. The greatest drawback, however, lay in the fact that the thickness of drum plates was limited to $2^3/_4$ inches because it was not practical to use rivets longer than required for this amount of thickness. This limitation in turn dictated maximum pressures for riveted boilers. Fusion welding overcame this severe drawback and led to important new designs and methods of construction.

In 1930, the United States Navy adopted a specification for construction of welded boiler drums for naval vessels, which were soon installed in a number of ships of the line. The rules were not unlike those that had been written by the Boiler and Pressure Vessel Code Committee for the fusion welding of drums for power boilers. The Brown Paper Company of Monroe, Louisiana, was recorded as having purchased the first power boiler drum constructed under this new Code.

An important ongoing trend was the planning, design, and construction of medium-sized industrial power plants and, with it, the demand for specialized boilers and other power equipment. Changes were also taking place steadily in the use of fuels, with oil and natural gas for the first time taking the ascendency over coal in a number of regions in the United States. This intensified interest created a demand for new types of boilers with medium-range capacities that would be suitable for power plants of this size, yet at the same time provide the efficiency and cost-saving advantages that had previously been available only in the large boilers found in central power stations.

For the first time — in terms of practical, commercial installations — power engineers were designing specialized types of boilers and furnaces, the objective being to meet special needs and at the same time assure efficiency and economy of operation. Each step toward specialization taxed the experience and dedication of those committees formed to provide codes and standards.

Marine developments were significant during the 1930s, despite the dampening effect of the Depression. During the decade following World War I, shipbuilding — merchant and naval alike — had been in recession. But by the 1930s, particularly by the middle of the decade, the industry became active, largely because of military developments in Europe, as well as the increasing demands of international trade prior to the advent of commercial planes capable of crossing the oceans.

The United States Navy stimulated the development of new types of marine boilers, with its demands for units that were not only more efficient and powerful, but more compact and economical. The weight factor was particularly critical because the Navy was still curtailed in its shipbuilding programs by the restraints of the Washington Naval Limitations Treaty.

"Out of these treaty limitations," explained a Navy spokesman, "came a serious challenge: to decrease fuel consumption on Navy vessels while at the same time increasing the cruising range. This demanded more boiler efficiency, of course, but also called for higher pressures and temperatures so that the turbines could deliver the ultimate in power."

Size was vital in many ships, notably in destroyers, frigates, escorts and other smaller vessels where quarters were congested and a great deal of equipment had to be stowed in cramped spaces. Many aspects of boiler and pressure vessel design for marine installations have been influenced over the years by the objectives, policies, and research of the Society of Naval Architects and Marine Engineers, which was organized in 1893, largely through the efforts of Rear Admiral Francis T. Bowles of the Construction Corps of the U.S. Navy.

The objectives were set forth in the Society's Constitution, "the most fundamental of which is the advancement of the art, science, and practice of naval architecture, shipbuilding, and marine engineering of all types in the United States."

Among the more recent developments cited were the entry of the steam turbine, electric drive, and diesel engine as competitors of the reciprocating steam engine, the development of the water-tube boiler, and the improvement of safety features in the design of ships.

In 1936, the Society sponsored the first International Meeting of Naval Architects and Marine Engineers.

In the late 1920s and early 1930s, the welding process of vessel fabrication came on the scene. This made possible the quantum jump in pressure because it eliminated the low structural efficiency of the riveted joint. This was widely utilized by industry as it strove to increase operating efficiencies by the use of higher pressures and temperatures, all of which meant thick-walled vessels.

Welding was a phenomenon in that it had been around in various forms for a long time but suddenly, almost overnight, was coming into its own. How? And why?

In his detailed history of the ASME Boiler Code, Arthur M. Greene, Jr., referred to the period at the end of the 1920s and the beginning of the 1930s as "The Great Years." Why did he designate this era in such a way and what were his real interpretations of "Great?"

His first criterion, he commented, was the advent of autogenous welding (later called "fusion welding") which received final approval and widespread acceptance during the period 1928 to 1931. During this era, welding was meeting the long-lived expectations of designers and fabricators for the shells and components of boilers and pressure vessels.

From ancient times until the early part of the 20th century, the only form of welding in significant commercial use had been forge welding. In the late 1800s, the first patents were issued that covered the use of the electric arc for welding and at about the same time, the first resistance welding machine was produced. The use of the oxyacetylene flame for welding was introduced in the early 1900s, though with limited applications. It was not until after World War I that welding had any real impact on the fabrication of metals, and even at first it was not known for certain how reliable welds would be in the long term or when subjected to any number of exterior factors.

Arc welding soon comprised the largest single group of processes, of which manual shielded metal arc welding became the most widely used. In the field of gas welding, the oxyacetylene process became the most important. An important development took place in 1907 when a Swedish engineer, Oscar Kjellberg, patented a coating for use on the bare wire. The purpose was to protect the wire from air and gases that tended to make the weld brittle. However, this method and several others that were patented during the next two decades or so proved to be only partially effective. It was not until late in the 1920s that cellulose was introduced as a component in the coating. During the welding process, the gaseous products formed by the decomposition of the cellulose served as a shield to protect the molten metal from the surrounding air. Thus the weld could be completed without weakening the ductility of the metal.

In 1930, a process was developed whereby the molten metal in the arc stream could be protected by a mantle of inert gas, such as argon or helium. In early applications, a nonconsumable tungsten electrode was employed for striking the arc, while additional filler material in wire form was introduced, as necessary, to complete the joint. This method became widely used for welding aluminum, stainless steel, magnesium, and the many other modern metals that increasingly came into demand.

As welding processes became more effective, reliable, and sophisticated, they gave birth to a whole new field of technology that soon required a knowledge of many associated fields of engineering. There evolved a surprising number of methods for welding, springing into demand because of the increase in types of applications and the development of new alloys not previously available.

It became clear to the Boiler and Pressure Vessel Code Committee and various subcommittees whose work involved the study of welding and welding processes during the 1930s that the formation of codes and standards in this field required specialized knowledge and would become more and more demanding as time went by. The multitude of factors involved in the proper application of welding became increasingly meaningful as the specification tolerances became more stringent. Those concerned with quality control were aware that they had to know and recognize many of the elements that were demanded in production fabrication, including the service requirements of the materials and products in question.

Fundamental to any plan of action was an engineering review of the structure, whether a boiler, pressure vessel, or other unit of equipment that was to be welded in order to determine the magnitude, nature, and distribution of the loads or stresses to be carried. It was also critical to be aware of other factors that might influence fabrication, installation, or stability, such as high heat, low temperatures, or chemicals. It was mandatory also to review environmental situations that would be encountered and which could possibly precipitate rusting, corrosion, or other deterioration or uncommon strain in the locations specified for operations.

Out of these fundamental studies and reviews there evolved a whole new technology of examination and testing that could be applied to welding techniques and practices and the materials and products that had been welded. In addition to visual inspection and other rudimentary methods that had been used for years, there came into being much more sophisticated nondestructive tests, such as the use of X-rays or other forms of radiation and the use of water or pressure applied by hydraulic means.

It was evident to the Code Committee and to relevant subcommittees that boilers and pressure vessels of all kinds represented the most demanding of all the many fields in which welding was playing a significant part. There could be little flexibility in the codification of tolerances, standards, and specifications. The initial work of the Welding Research Council was a vital factor in the acceptance of welding and in the improvement of procedures and equipment. It should be recognized here that WRC and various other organizations involved in specialized research and development were motivated by, and with the full support of, ASME. Such organizations came into being because, while it was critical for the Code Committee to compile technical data continuously, the Committee was only involved with the evaluation of research and not the function of research per se.

Among the projects researched by WRC were many that related directly to the work of the ASME Boiler and Pressure Vessel Code committees. Representative research included:

— the effect of high temperatures and pressures on steels used in pressure vessels;

— control of distortion and shrinkage in the welding of boilers and other units;

— the effect of internal pressure on nozzle connections and pressure vessel heads.

A Pressure Vessel Research Committee was established by the American Welding Society in 1945, after about a decade of increasing interest and research in this field. One of the underlying objectives was to study problems, on a cooperative basis, "to provide information on which the ASME Boiler and Pressure Vessel Code could be expanded and improved. A common idea was that the research should have a practical approach and a strong chance of producing tangible results within a reasonable period of time."

The research program started out with the focus on problems in four key areas: (1) materials, as related to their service suitability and as affected by production methods; (2) safer designs at equal or greater economy; (3) techniques of fabrication to avoid damage to materials, and (4) the inspection and tests of materials, fabrication practices, and completed vessels.

Some of the questions that had arisen during the 1930s and had never been fully answered included the following, among others.

- How can impact tests or other tests on laboratory specimens be tied in with the performance of pressure vessels?
- How much ductility in base metal and weld metal is necessary for the safe operation of pressure vessels in various services?
- What is the effect of tolerances on maximum allowable stresses in pressure vessels?

- What degree of weld defects does not affect the safety of a vessel?
- What is the purpose of a hydrostatic test?
- What are the relationships of temperature, character of loading, and restraint to failure in a vessel?
- What is the effect of temperature gradients on stress distribution in thick-walled vessels?
-

Other questions focused on the heat treatment of carbon steels, maximum working stresses around openings and junctions, formulas for safety at high temperatures, and some adverse conditions that lead to age-hardening and strain embrittlement.

The Pressure Vessel Research Committee was made up of numerous subcommittees, including those whose work related specifically to pressure vessel steels; hydrogen embrittlement; the nondestructive examination of materials; bolted flange connections; elevated-temperature design; stress indices for piping, pumps, and valves; welding procedures; thermal and mechanical treatments; and other such topics.

The subject of inspection policies and procedures became more important than ever with the advent and spread of welding methods and techniques. One of the most important functions of an authorized inspector was to examine welding procedures and operators for their compliance with the intent of the Code. While the Code did not dictate methods to the manufacturer, it did require him to prove that his welding was sound and that his operators were qualified, as set forth in Section IX, Welding and Brazing Qualifications.

As welding procedures and operations became more of a factor in the Code, it became necessary to orient those involved with the nature of welding technology as the study of the properties of welded metals and the thermal and mechanical effects that transpired during the processing of metals. These included studies of the movement of the atoms in a liquid state and the formation of the grain structure of the metal when changing from the liquid to the solid state. This understanding of the behavior of metal during welding was vital to any establishment or modification of codes and rules relating to the design and fabrication of boilers and pressure vessels in order to prevent dangerous residual stresses or defects in the welded joints.

Metallurgists in this field came to think of metals in terms of a "life cycle," in which a metal was "born" during a melting operation, suffered "diseases" caused by stress, pressure, temperature, chemicals, and other factors, and eventually "died" when it failed during service or was purposely scrapped. The effects of various forms of heat treatments became the subject of a considerable family of research, experimentation, and testing, as projects were undertaken to determine how metals could be hardened, softened, toughened, or made more ductile.

Tests were also conducted to determine why and how some metals were much more difficult to weld than others, and what part was played by chemicals, metallurgical changes, the heat of the arc, the nature of the flux, and other factors. These revealed the manner in which the atoms, in the metals used, arranged and rearranged themselves in regular patterns and grain sizes as the cooling process began and ultimately was completed. In one such research project, for example, an accomplished consulting metallurgist made tests and prepared a pair of charts to show the differences between ferritic (carbon and low-alloy steels) and austenitic (nonmagnetic chromium nickel steels) materials.

Such research and testing played an important role in establishing codes and specifications in the early 1930s when welding was strongly coming into its own as a replacement for riveting in major fabrications and installations, and when research was undertaken to determine stress factors and relief.

Code requirements for stress relieving in boilers and pressure vessels that have been welded have become familiar today. But in the late 1920s and even through much of the 1930s, stress relief was still something of a mystery. In part, the paucity of data can be traced to the lack of testing equipment that was sophisticated enough to study and compare samples of metals and welds before and after stress relieving was applied. It was little realized that certain conditions affected the degree of stress, such as the thickness of a welded joint, the effect of low temperature on a vessel, or the changes that would take place under high pressures. In many cases, proper examination, testing, and adjusting demanded far more time-consuming efforts than were originally intended.

One significant event at the start of the 1930s was the Twelfth National Metal Exposition in Chicago at which five societies prominent in the metallurgical field held their meetings, namely the American Society of Mechanical Engineers, the American Welding Society, the American Society for Steel Treating, the Institute of Mining and Metallurgical Engineers, and the Institute of Metals.

"Only a few years ago," reported a release from the Exposition, "welding was looked upon as a means of making joints which were not supposed to be subjected to dangerous stresses, and the failure of which would not involve fatal consequences to life and limb. The attitude toward welding then was not only highly critical but in some quarters quite unfriendly. But the art kept on improving and a large variety of welding methods, particularly electric, were developed, such as the arc method, the quasi-arc, and the many forms of resistance welding."

99

"Rapid advances" were mentioned, too, in such welding supplies as rods and fluxes, in automatic welding machinery, in inspection methods, and in standardization. These factors also led to a decrease in the cost of operation. By this time, the stage had been reached in the manufacture of a welded product at which "the weld may be assumed to be as strong as the base metal and to be capable of withstanding stresses which only a few years ago would have been considered unbelievable."

One of the significant events at the Metal Exposition, along with exhibits of welding machines and materials, was the reading of a technical paper describing a nondestructive method of magnetically testing butt welds. "The value of this paper lies in the fact that even in automatic welding each weld stands by itself, and only when reliable methods of nondestructive testing have been developed will it become possible to use welds with absolute certainty as to their strength and reliability." It was evident that nondestructive testing was, or soon would be, an essential element in the ASME Boiler and Pressure Vessel Code.

CHAPTER 11

PRESSURE VESSEL CODE DEVELOPMENT

New Products and Materials Challenge the Code

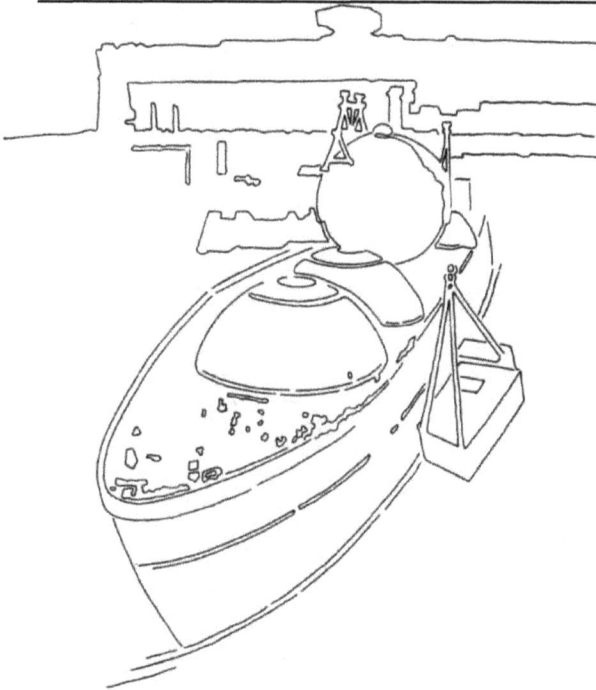

To the Boiler Code Committee - Gentlemen: The object of the code covering the construction of Unfired Pressure Vessels is to provide safe construction for the great variety of pressure containers now manufactured and used for innumerable purposes throughout the country and under an exceedingly wide range of conditions. The efforts to give publicity which have been made in order to obtain as wide a range of opinion as possible, have brought out views which were often times diametrically opposed. In such cases, the attempt has been made to frame the rules in accordance with the greater weight of opinion."

Thus began the letter of introduction to Rules for the Construction of Unfired Pressure Vessels, Section VIII, Unfired Pressure Vessel Code, dated 1925. The letter stated that the rules would not only result in safer vessels but would eliminate failures that could take place because of a lack of restrictions governing materials, workmanship, and supervision. The rules covered safety appliances; the thickness of pressure vessels carrying corrosive substances; hydrostatic tests; examination of vessels by qualified inspectors; the internal examination of pressure vessels fabricated in whole or in part by welding; and the authentic stamping of pressure vessels with ASME data and the name of the manufacturer. The 1925 Edition of Section VIII also contained rules for various processes of welding. These covered, in order:

- *Fusion Welding*-with a discussion of the processes (either oxyacetylene or electric-arc); definition of the term "base metal"; techniques and methods of welding; hydrostatic and hammer tests; and guidelines for correcting defective welds and distortions

- *Forge Welding*- with attention to the thickness of shell plates; heating agents and temperatures applied during fabrication; the lapping of welds; inlet and outlet connections; and annealing to remove internal strains and distortions

- *Brazing*- with a description of materials and thicknesses; the reinforcement of threaded openings; the lapping of longitudinal seams during brazing; and the correct method for driving head seams into the shells with a tight, driving fit before being brazed

- *Enameled Vessels* -with a description of their design and the nature of the enamel or glass used for coatings and linings; single-shell and double-shell (or jacketed) types; the extra factor of pressure on inner and outer shells; welding methods and restrictions; and the necessity of re-rolling cylinders after welding

An early revision of the first edition of the Unfired Pressure Vessel Code included a list of allowable stresses for steel at elevated temperatures. Several petroleum engineers pointed out that these were lower than had been in successful use for several years by the petroleum industry for the kinds of pressure vessels that were designed for refineries. With the temperatures for petroleum processing gradually increasing, some of the oil companies had been conducting in-depth research on the creep characteristics of steel at elevated temperatures.

Charles S. Haupt, vice president and chief engineer of the Standard Oil Development Company and chairman of the API Committee on Refinery Equipment, was asked to appear before the ASME Boiler and Pressure Vessel Committee to explain and justify the successful and safe use of higher stresses in pressure vessels used for petroleum processing. Accompanied by Walter Samans, chief engineer of Atlantic Refining Company, Haupt appeared before the Committee and requested that ASME revise its stresses upward. In reply, the Committee explained that the higher temperature stresses had been established by reducing their allowable stresses at normal temperatures by specific percentages rather than on the basis of any scientific knowledge. It was further explained that the ASME Code had been written primarily for vessels designed for compressed air storage and similar uses, and not for vessels designed for the processing of petroleum or chemicals.

The net result was that the ASME Code Committee agreed to exempt the latter types of vessels, since the petroleum industry desired this change and as long as the API would issue a pressure vessel code of its own. Haupt and Samans accepted this stipulation and promised that the API would indeed establish a committee to prepare the industry's own modification of the Code.

When the API committee went into session, one of its objectives was to modify what its members felt were overconservative features in the Code, as represented by the extra safety factor provided by its factor of five. The factor of safety of four in the API code was less conservative and, in effect, permitted the use of thinner vessels than with the factor of five. This judgment was based in part on the improved strength and capabilities of contemporary materials as compared with those of the past.

The ASME Code Committee was kept regularly informed about the program and progress of the API Committee. As the code neared completion, the ASME committee decided that it would be better if it were identified in some fashion with the new code. Thus it was that the Joint API-ASME Committee on Unfired Pressure Vessels was appointed in December 1931, composed of four members from each group, with the objective of preparing "a code for safe practice in the design, construction, inspection, and repair of unfired pressure vessels for petroleum liquids and gases." The first edition was published in 1934.

"The men on those committees were the shapers of the early pressure-vessel code," explained Leonard P. Zick. "The refineries were self-contained and were not perceived to have public exposure. However, the wide variety of processes required a far greater range of materials, designs, and methods of construction than had been the case for boilers.

"Processes for refining gasoline had many phases of development, but always required large pressure vessels. For example, the cat crackers were designed for relatively low pressures and temperatures and were built of carbon steels. Later, the hydrocrackers required the development of improved properties of heavy chrome moly low alloy plate. The process design was limited by the elevated temperature properties achieved in the heavy plate."

As he pointed out, since gasoline was stored and handled at atmospheric temperature and pressure, its containment did not require pressure vessels. However pressure vessels were required for the storage and transportation of volatile gases such as propane, butane, ethylene, ammonia, carbon dioxide, oxygen, methane, nitrogen, and hydrogen.

"The economics of storage," he added, "depends upon the pressure and maximum temperature to which the pressure vessel will be exposed. When liquefied, they may be stored at atmospheric temperature; however, even liquefied gases are sometimes stored in insulated pressure vessels. Examples are the storage vessels used at test or launch facilities for liquid-propelled rockets."

Among the most important types of pressure vessels in the petroleum industry that had to be studied and understood by Code committees were those used for the transportation and storage of Liquefied Petroleum Gas. This product, referred to familiarly as LPG, was composed predominantly of propylene, propane, butanes, and butylenes, is a commodity that, while much in demand today, was formerly just a waste byproduct of the petroleum industry. For many years, the only practical way of transporting LPG was in cylinders manufactured according to the specifications of the Interstate Commerce Commission (ICC), which later became the Department of Transportation. The pressure relieving devices used on these cylinders had to be approved by the ICC as to the type and size, based on fire and other tests.

From its inception, the API-ASME Code contained a chapter designated as Section I that covered recommended practices for inspection and repair and for establishing allowable working pressures for vessels in service. A special feature of this chapter was its recognition and provision of ratings and bases for handling problems of vessels subject to corrosion. Although Section I, like other parts of the API-ASME Code, originally was intended for vessels used in the petroleum industry, the provisions were considered applicable to pressure vessels in most services.

In the 1920s, the ICC approved the experimental use of railroad tank cars that had been specially fabricated by the use of fusion welding to transport LPG. For the first time, there was an economic means for transporting this volatile product to large and distant markets. The perfection of this system increased the demand for containers that were larger than the traditional cylinders and that could be used for storage at industrial, commercial, and residential sites.

With this increased use and movement of LPG, it became apparent to the petroleum and chemical industries that strict standards would have to be established to ensure the safe transportation, storage, and use of the gas. The National Fire Protection Association (NFPA) agreed to the formation of a national committee to review the requirements and publish the necessary codes and standards. The end result was NFPA Pamphlet No. 58, issued in 1932, "Standards for the Storage and Handling of Liquefied Gases," later adopted in whole or in part by all states. The rules for fabrication included riveting, fusion welding, forge welding, and brazing. Rules were provided, too, for unreinforced openings and for reinforcement of openings.

The 1935 Edition included nonferrous tubes and cast steel as permissible materials. Other changes involved new rules for vessels under external pressure, bolted flanged connections, flat heads, and electric-resistance butt welding. During the rest of the 1930s, the most significant revisions had to do with containers for gases and liquids at very low temperatures; the first reference to impact tests in the Code; the first sanction of a safety factor of four (permitted in this case for seamless construction, accompanied by a number of restrictions); the first appearance of rules for the qualification of welding processes and operators; and the extension of stress values for some steels to temperatures up to 1,000°F.

During this period, the foreword to the Code book became standardized and no longer presented a resume of the main developments that had occurred since the previous edition. This reflected the growing number of projects underway, the increasing length and complexity of revisions, and the need to conserve space.

From the beginning, the containers used for storage and handling LPG, other than ICC cylinders, had to be constructed in accordance with the requirements of the ASME Boiler and Pressure Vessel Code under Section VIII, Unfired Pressure Vessels, Division 1, or the joint code of API-ASME on unfired pressure vessels. The safety relief valves were set to discharge at a design pressure of the container based on a safety factor of *four*. While this was in agreement with the API-ASME Code, it did not comply with the ASME Code.

The rules required the LPG industry to determine the flow of safety valves in accordance with a formula created by John F. Fetterly, then an inspector for the Bureau of Explosives of the Association of American Railroads. The formula, which was published in the Bureau's Annual Report, was so reliable that it remained in use with no revisions until the 1950s when some minor modifications were made.

By 1939, the API-ASME Code had become widely recognized, with a few exceptions. Starting in 1947, the LPG industry required that the flow capacity of each type, size, and pressure setting of their safety relief valves, instead of calculating their flow from physical measurement, be determined by test. Prior to the issuing of the 1950 supplement of Section VIII, the Subcommittee on Safety Valve Requirements of the ASME Boiler and Pressure Vessel Committee recommended that, for fire conditions, pressure relieving devices be allowed to permit the pressure in a vessel to raise to 120 percent of the design pressure. This corresponded to the practice of the LPG industry for the previous 20 years and was in addition to the requirement of a maximum pressure rise in the vessel for operating at 110 percent of the design pressure of the vessel. ASME still required that the safety valve be rated at 110 percent of its set pressure for flow capacity. LPG valves were flow tested at 120 percent. In 1950, also, the LPG industry specified that its safety relief valves be stamped according to its requirements, and according to Section VIII of the ASME Code.

Because of the possible hazard of escaping gases, the valves used by the industry had soft seats or sealing rings, which eliminated any leakage until the valve itself opened in the event of an emergency. Since the valves, therefore, remained closed over long periods of time, it was discovered that these periods of inaction caused the soft seats and rings to become compressed. Even though the depression that resulted was ever so slight, its effect was to lower the start-to-discharge pressure. To compensate for this action, the tolerance for setting the valve's start-to-discharge pressure was therefore established in 1955 as 0 percent to 10 percent of design pressure.

As a result of the discontinuance of the API-ASME Code for Unfired Pressure Vessels in 1956, a demand arose for the issuance of its Section I as a separate publication, applicable not only to vessels built in accordance with any edition of the API-ASME Code, but also to vessels built in accordance with any edition of Section VIII of the ASME Boiler and Pressure Vessel Code. Such a publication seemed in order for industry to have assurance of preserving the same trend toward uniform maintenance and inspection practices that had been afforded by Section I of the API-ASME Code.

The Subcommittee on Safety Valve Requirements of the ASME Boiler and Pressure Vessel Committee began, in 1974, to review the differences between the specifications in the API's Pamphlet No. 58 edition and those specified in ASME's Section VIII. **Most** were easily reconciled. One major difference concerned the final rating of the valves after the mandatory flow tests. ASME required that the average flow quantities obtained from the testing of three samples of each combination of design, size, and test pressure should be reduced by 10 percent to obtain the official rating. The LPG industry specified that each design, each size, and each pressure rating be tested, instead of obtaining coefficients of discharge (as required by ASME) and applying the value secured from a small group of valves to all in that class.

Since the values obtained by testing *each* design, size, and pressure setting provided a true value for this type of valve, it was therefore not necessary to make the 10 percent reduction specified by ASME.

Except for these differences, the two codes remained practically the same right into the 1980s.

Some 17 years after its founding the Pressure Vessel Research Committee, with financial support from other organizations like ASME, took a step that proved to be very useful to the Boiler and Pressure Vessel Committee and other bodies charged with the establishment of codes and standards. This was the formation of the Program Evaluation Committee, whose basic responsibility was to study all of the research programs in progress and then summarize and present results in a condensed format. This not only brought the findings into sharper focus for Code committees, but greatly reduced the time and effort required to reach decisions about rules and specifications.

Eventually the Program Evaluation Committee also was to play a role in determining the most significant problems for future study. Advance information about forthcoming programs was valued because it alerted Code committees in an orderly manner to subject areas that they would have to anticipate and consider in the near or distant future. The first release was a bulletin containing a list of no fewer than 35 such problems, published and distributed under the title, "Long-Range Plan for Pressure Vessel Research."

Characteristic of the research programs undertaken and completed by the PVRC were the following.

- The influence of warm prestressing on the strength, ductility, and toughness of pressure vessel steels, supported by tensile, impact, and low-temperature fracture toughness evaluation tests.
- Determination of the mechanical properties of heavy-section steel plates, weldments, forgings, and castings and the reliability of

nondestructive examination methods for discontinuities in such components.

- Effectiveness of computer analysis of the design of spherical and ellipsoidal heads on pressure vessels in determining types and degrees of stress.

- The development of more effective procedures for the design of un-symmetrical flange combinations, integral type flanges, and full-faced flanges, all of which resulted in specific recommendations for ASME Code adoption.

- Study of the effects of preheating, postweld stress relief, and electrode type on the notch toughness of welded joints.

- Test program for the design of heads against internal and external heads.

- Experimental investigation of radial and nonradial nozzles attached to spherical shells at different angles to establish the effect of the specific angles on the stresses at the welded connection.

- The behavior of pressure vessel steels in the temperature range from 700°F to 1,100°F, defining the specific temperatures at which creep-rate properties, rather than short-term elevated-temperature tensile properties, controlled the design stress for these steels. The study also provided data on the high-temperature creep-rupture strength and ductility of the heat affected zone and weld metal of typical weldments of the units tested.

The data from research programs like these proved to be invaluable throughout the years in the planning and publication of the ASME Boiler and Pressure Vessel Code.

Taking a backward look, it was evident that pressure vessels had come a long way from the time at the end of the 1930s when the Code Committee had painfully acknowledged that "all too frequently, combustion explosions occur in industrial pressure vessels, particularly compressed-air receivers, resulting in loss of life and property damage." The research undertaken to correct this problem, late though it was in being initiated, had represented "the first step in determining means to prevent this loss of life and property damage when combustion explosions do occur in pressure vessels by protecting the vessels from excessive overpressure with suitable frangible rupture disks."

In many industrial applications of the day, pressure vessels were being used for the storage of potentially explosive mixtures of compressed air and some kind of combustible vapor. A common example was the air receiver. Even though compressed air was not "combustible" in itself, it became so if there was any kind of oil leak into the chamber, either through a defect or faulty operation. Unexpectedly, air receivers were causing many casualties.

Research showed that relief valves, though adequate to protect such vessels from overpressure, could not compensate for the extremely rapid pressure rise that occurred when combustion explosions occurred. Ironically, although a great deal of research had been conducted on the explosive forces of small bombs, for example, almost nothing had been done in the field of larger closed storage vessels. One of the positive results of the studies that finally did take place had been the design of inexpensive rupture disks that either prevented or greatly reduced explosions of this nature.

During the late 1930s and early 1940s, one of the active ASME groups was the Special Research Committee on Vessels Under External Pressure. By the time of the publication of the 1943 Edition of the Code, this committee had amassed a wealth of useful data on pressure vessels, which ultimately saw its way into the Code in one form or another. There were rules, for example, that applied to at least 20 different types and specifications of steels. Within the next three years, this single subject was expanded greatly, with the addition of a still greater range of steels for use at both ordinary temperatures and elevated temperatures, and under widely varying conditions. The work of the committee resulted also in many other additions and revisions, including new charts for nickel and aluminum alloys, rules for the reinforcement of openings, the testing of vessels, and odd-shaped vessels designed for unique purposes.

The Code Committee was constantly faced with the bilateral challenge of trying to include new technological data and yet, at the same time, simplifying the process of references and interpretation. "As the need for rules to cover steel vessels at elevated temperatures became greater," explained one member of the committee, V. F. Hartman, "a search was made to determine a practical method of applying correction factors to the chart to adjust for elevated temperatures."

However, as was increasingly the case year after year, even the most imaginative chart design failed to cover all of the necessary points and it became necessary to publish *two* charts in a section where one had formerly been sufficient. This was a trend that could seldom be avoided.

The trend toward complexity could be seen in the rules covering just about every type and kind of individual element as designs were modified or new products came into being. The safety valve is a good example.

Over the years, three different types of valves were developed, each one a modification of the other two. Thus the Code defined a *safety valve* as "an automatic pressure relieving device actuated by the static pressure upstream of the valve and characterized by full opening pop action." It was specified for use mainly for gas or vapor service.

A *relief valve* was defined as "an automatic pressure relieving device actuated by the static pressure upstream of the device which opens further with the increase in pressure over the opening pressure." It was identified mainly with liquid service.

A *safety relief* valve was defined as "an automatic pressure relieving device suitable for use either as a safety or relief valve, depending on application."

Changes in the Code for these valves were minimal during the 1930s, mainly specifications that kept pace with the contemporary increase in pressures, temperatures, and other factors that affected safety. By the mid-1940s, safety and relief valves "were sized for the first time by actual flow test based on the diameter and length of the container." In 1950, "for the first time where the excess pressure was caused by exposure to fire or other sources of external heat, the relieving capacity had to be sufficient to prevent the pressure from rising more than 20 percent above the maximum allowable working pressure of the vessels when all other pressure relieving devices are blowing." In 1953, "the safety relief requirements were for the first time based on the surface area of the container and not on the container diameter and its length."

Despite the intricacies of Code specifications and the continuing effort to provide uniform rules and standards, there were limitations in what any code could accomplish when a vessel had been installed and was in practical use. "Two new vessels built for equal pressures and temperatures and the same type of service can fare very differently in different locations," wrote Robert Chuse, an expert on the subject, in a handbook published in 1960. "In one plant, capable maintenance personnel will keep the vessel in top shape by means of a systematic schedule of inspection and maintenance. In another, an equal number of personnel, through ignorance, incompetence, or neglect, can bring about the failure of the vessel in a comparatively short time. Competent personnel are obviously key factors in the longevity and safety of a pressure vessel."

Pressure vessel history showed that a continuing problem was corrosion, a subject to which the Committee addressed itself almost from the first Pressure Vessel Code publication. Many types of unfired pressure vessels were found to be subject to deterioration by corrosion, erosion, or both. Corrosion occurred in the form of pitting or grooving, in some cases locally and in others over large areas - the latter having the effect of seriously reducing the plate thickness. Yet Corrosion defied inclusion directly in the Code because its occurrence or prevention depended upon situations and factors that could not be specified by rules and that were really beyond the jurisdiction of the Code itself. Nevertheless, the Committee pinpointed one generic weakness: some designers were not being farsighted enough to anticipate flaws in materials and equipment that could develop from long-term usage.

Stress failures were also a prevalent cause of problems during the early years because many plate materials were not really appropriate for the service and durability expected of them. In other cases, the materials were found to be basically suitable, but would begin to fail, sometimes rapidly, under stressed conditions. One example of this kind of failure was that of a chemical company which had installed stainless steel corrosion buttons as a means of measuring corrosion in one of its large fired cast-steel vessels. When this vessel ultimately had to be replaced, the company installed a new unit that had been fabricated from the same stainless steel used for the buttons. It was, with good reason, expected that this was a wise move because the buttons had demonstrated a high degree of corrosion resistance.

After the vessel had undergone a few days of service, however, the operator noticed leakage through the 1 inch shell, reduced the pressure immediately, and shut down the operation. Examination revealed, to everyone's surprise, multiple cracks in the stainless steel. As it later turned out, the substance being processed was highly corrosive to the grain boundaries of the metal when under stress. When a test was made under both stressed and unstressed conditions in the same solution, the stressed parts failed within 100 hours.

It was discovered, too, that excessive stresses could develop in vessels that were not free to expand and contract during regular changes in temperature. When stresses like these were permitted to occur with too much frequency, the users could expect eventual failure from fatigue.

Code rules for inservice inspection became increasingly important when it was discovered - sometimes through bitter trial and error - that vessels were installed in locations that actually prevented or seriously impeded examination through the openings provided for such purpose. Such situations were in direct contradiction to rules requiring inspection openings in all vessels subject to internal corrosion, erosion, or mechanical abrasion - conditions that, in fact, applied to almost all vessels in one form or another. Reviews of this element of risk led to the official caution that users bear the responsibility for selecting only those vessels whose designs permit proper access for inspections after the vessels have been installed.

The year 1972 saw publication by ASME of a reference work entitled *Pressure Vessels and Piping: Designs and Analysis*, subtitled A *Decade of Progress*. This monumental two-volume compendium was published as a sequel to an earlier edition, which contained technical papers that had been written and published between 1927 and 1959.

Volume One covered Analysis and was dedicated to Bernard F. Langer at the time of his retirement from Westinghouse. The compilation of materials, explained the five editors in charge of the project, "makes conveniently available many technical papers that have contributed to the development of the Codes." The objective, as described, was "to summarize the state of the art and recent advances in the many relevant topics in pressure vessels and piping design; to collect in one book selected papers and bibliography; to provide technical background for the many users of the ASME Boiler and Pressure Vessel Code; and to contribute to the ability of the young engineer to be more effective."

The editors, following long-standing ASME policy, cautioned that "under no circumstances, however, should any of the material in this volume be construed as a supplemental Code requirement."

It was emphasized that the literature on pressure vessels and piping alone had grown at "an awesome rate ... since the Code has been changed, enlarged, and improved at a pace consistent with the rapid advance in the technology," making it difficult for engineers and others in this field to select and review technical papers of greatest concern to them. The book was divided into three distinct parts, covering Analysis for Design, Structural Components, and Structural Dynamics.

The Design section covered such topics as the general range of design problems in regard to the ASME Code; profiles of the key professional organizations that have contributed to the advancement of pressure vessel technology; a comprehensive and informative review of the whole subject of fatigue; and a survey of computer programs available for the static, dynamic, and nonlinear analysis of shells and shell type structures.

The section on Structural Components dealt with the buckling of cylindrical shell end closures through internal pressure; a pioneering example of photoelastic analysis when applied to the design of flanges; a review of the first comprehensive analytical treatment of a piping system, which was still being referred to and found valuable by engineers; the effect of internal pressure on the flexibility and stress intensification factors of curved pipe or welding elbows; design guidance for an often needed and relatively simplified method of supporting horizontal tanks; and many other subjects of this nature.

The editorial selections on structural included, among other projects, a seismic analysis of primary piping designs and components for nuclear generating systems; evaluating and predicting the vibration response of piping induced by random pressure fluctuations in turbulent, axial flow; the dynamic stability of cylindrical shells subjected to sonic waves because of a loss-of-coolant reactor accidents; and a method of solution for a common problem facing designers of nuclear reactor components: the impact of components with one another under dynamic loading conditions.

Summing it all up, the book reminded readers that "in the last decade, tremendous change has occurred in pressure vessel design criteria for metal temperature below 800°F. The present design criteria were a direct consequence of a better understanding of material behavior, such as low-cycle fatigue, brittle fracture, plasticity, mechanical and thermal ratcheting, and shakedown, as well as the progress made in electronic computer and related computational methods .... As a result of this new approach, today's Code design criteria are more consistent in overall safety and reliability."

An important international milestone in the history of codes relating to pressure vessels was the First International Conference on Pressure Vessel Technology, held in Delft, Holland, in the fall of 1969. Historically, said a report on the proceedings, during the evolution of boiler, piping, and pressure vessel codes, technical groups have set their sights on sets of rules that would assure public safety and reflect the many economic and technological factors of their environment. By adopting this goal, the codes have provided protection against several different failure modes that must be considered by the users, designers, and fabricators of pressure vessels. The rules, while covering various categories and types of stresses, also gradually and productively have presented data for control and protection against the failures that had been defined.

During the evolution of these codes, the technical groups involved in their preparation had traditionally been composed of designers, manufacturers, users, and representative specialists from associated industries and the academic world. Jointly, they determined that there were three critical links upon which code criteria could be built: failure modes; stress categories and stress limits; and material considerations.

As technology expanded and design requirements became more stringent and specialized, it was found that closer attention had to be given to failure modes and the protection which the codes insured against them. In dealing with modes of failure, designers now had at their disposal complete data on the materials and parts in question. Yet they still had to ask themselves, Code criteria cautioned, what the numbers meant in relation to the adequacy of the design. Would they insure safe and satisfactory performance of a component? It was against these pressure modes that the designer had to compare and interpret stress values.

For example, elastic deformation and buckling could not be controlled by imposing upper limits to the calculated stress alone. Designers also had to consider the geometry and physical stiffness of a component, as well as properties of material. In addition to setting values for allowable stress, the designer also had to devise an adequate failure theory in order to define how the various stresses in a component would combine, react, and contribute to the strength of that part.

In the matter of *stress categories* and *stress limits*, it became apparent that consideration had to be given to the different types of stress, because not all required the same limitations in order to provide protection against failure. To start with, there was *primary stress*, developed by the imposed loading necessary to satisfy the laws of equilibrium of external forces and moments, but not self-limiting. The limits specified were intended to prevent any plastic deformation and provide a nominal factor of safety on the ductile burst pressure, as well as control of creep and stress rupture.

Next under consideration had to be *secondary stress*, which was the type developed by the self-constraint of the structure. The basic consideration of secondary stress was that it was self-limiting, since minor distortions could satisfy the discontinuity or thermal expansions that caused the stress to occur. The combination of primary/secondary stress limitations were intended to control excessive plastic deformation and the possibility of incremental collapse.

The third category of stress, *peak* or *total stress*, was defined as "the highest stress in the region under consideration, the basic characteristic of which is that it causes no significant distortion." This type of stress was objectionable as a possible source of fatigue failure, corrosion, or brittle fracture. Peak stress limitations were formulated to protect components against fatigue failures that could result from cyclic loading, as well as the longer range problems of corrosion and embrittlement.

The third link upon which Code criteria were built, *material considerations*, was concerned basically with the various mechanical properties of steels and the possible modes of failure in vessels fabricated of these metals. The standard tension test was designed to provide data on yield strength,* tensile strength, and elongation characteristics, which data had been known and used by Code-writing bodies for many years. It had been discovered that various steels used for pressure vessels behaved quite differently when loaded beyond the yield point.

Austenitic stainless steels, for example, show a relatively low yield point, but require a substantial increase in stress to cause a significant increase in strain. These materials also show large ductility before failure. They were thus able, through a characteristic then described as "strain hardening," to redistribute the stress within a member in such a way as to allow more of the total thickness of the material to support the imposed load.

---

*Defined in the mid - 1930s as "the stress necessary to produce a given inelastic strain in a material" of the material to support the imposed load.

Contemporary research demonstrated that carbon and low-alloy steels of conventional pressure vessel grades also displayed some of the characteristics of strain hardening, yet to a lesser extent than the austenitic steels. At the other end of the scale were some of the the quenched and tempered steel alloys, which displayed relatively high yield strength, but showed little or no tendency to strain harden. Quite the contrary, they might soften under strain.

In all of these studies and reviews of the rules, the Code Committee continued to recognize the fact that stringent protection against brittle fracture was "a necessary consideration in any design." The Code afforded this kind of protection, indirectly, by controlling the allowable stresses on the basis of both yield strength and tensile strength. Thus, materials with high-yield/tensile ratios, which were more prone to brittle fracture, were provided with design stresses limited by a fraction of tensile strength.

Studies of ferrous materials and high-temperature alloys went hand in hand with many of the developments covered in preceding pages. During the 1950s, the ferrous materials activities of the Code Committee were reorganized to include high-temperature alloys, plates, tubular products, castings, forgings and bolting, notch sensitivity, and metals engineering. In the field of alloys, the Committee also reviewed carbon and low-alloy steels, high-alloy stainless steels, and superalloys, both ferrous and non-ferrous. The consideration of materials covered metallurgical and physical properties, specifications for products, the setting of allowable stresses, the strength of materials at both elevated and subzero temperatures, and the overall suitability of metals and other materials for use with the Code.

In selecting the membership of new subcommittees and subgroups, the long-established ASME procedure was followed of enlisting the aid of the best engineering talent in the nation, with proper representation from materials producers, manufacturers of boilers and pressure vessels, and users of the equipment. As in the past, one of the major functions of these groups was to help interpret the rules for construction of boilers and pressure vessels and to make recommendations for action by the Main Committee, as well as to draft revisions of the Code rules to keep up to date with technological progress, both in regard to design stresses and to new materials.

"The rules have long since proved their efficacy from the standpoint of safety, and standardization of the rules has greatly facilitated the construction and inspection of boilers and pressure vessels," reported the Code Committee at a meeting in Kansas City in 1961. "However, the very success of these rules and their wide adoption by insurance underwriters and by many state safety codes has made it very difficult to change the rules to keep pace with engineering progress."

116

Over the years since the formation of earlier ASME subcommittees in this area of operations, a number of concerns and issues had periodically occupied the prolonged attention of the members, it was also related.

One issue that kept arising was the fact that the use of a factor of four on the tensile placed the Code at a considerable economic disadvantage in competition with European construction, wherethe tensile was notconsidered except possibly for high-strength materials having yield-to-ultimate ratios of 70 percent or more.

Another problem that kept cropping up was the basic Boiler Code concept that ferrous materials should only be used in their annealed or soft condition. The rules at that time did not recognize credit for the increase of strength produced by the heat treatment of steel, for example. The reasons for the negative attitude toward quenched and tempered material could only be surmised since its originators were no longer on the scene. The only explanations the Code committees could come up with was the old and outmoded assumption that heat treated materials were certain to lose their strength during prolonged exposure to elevated temperatures. It had been an accepted fact from the original drafting of the Code in 1914 until well into the 1930s that heat treating was "more of an art than a science."

Dependable data were difficult to obtain, since much of the early research was of the "seat of the pants" variety and mainly conducted on short-term bases. By the beginning of the 1940s, however, engineers had learned with a fair degree of accuracy and reliability how and why steel hardened and what the effects were on the properties of various alloys when subjected to heat treating or welding. Much had also been learned about the behavior of steel under stress at both elevated and subzero temperatures. So it was unanimously agreed that the time had come to recognize these facts and revise the Code accordingly.

Subcommittees that were given the responsibility of reviewing communications and inquiries from outside - a duty that almost all of them were charged with to a greater or lesser degree - were now facing a growing challenge: periods of from ten months to more than a year were no longer uncommon when reviewing an inquiry and making a recommendation. Because of the increasing complexity of materials, processes, installations, and other factors, it was almost impossible for any subcommittee to devise reliable recommendations singlehandedly. After thorough review of the data at hand, one committee would invariably have to refer the inquiry to at least one other committee - and usually several at a time. What was needed was a *multi-faceted* recommendation, based on the individual expertise of each group involved.

One step taken to improve the time factor was to increase the size of the subcommittees so that there would be a broader knowledge of each subject within the membership of the committee. It became commonplace for subcommittees to have to turn to associated groups like the PVRC or the WRC for information which, in turn, very frequently triggered research or development programs that might take two or three years to complete and come up with applicable answers.

CHAPTER 12

THE TECHNOLOGY OF EXAMINATION AND TESTING

Science Catches Up on the Inspection Line

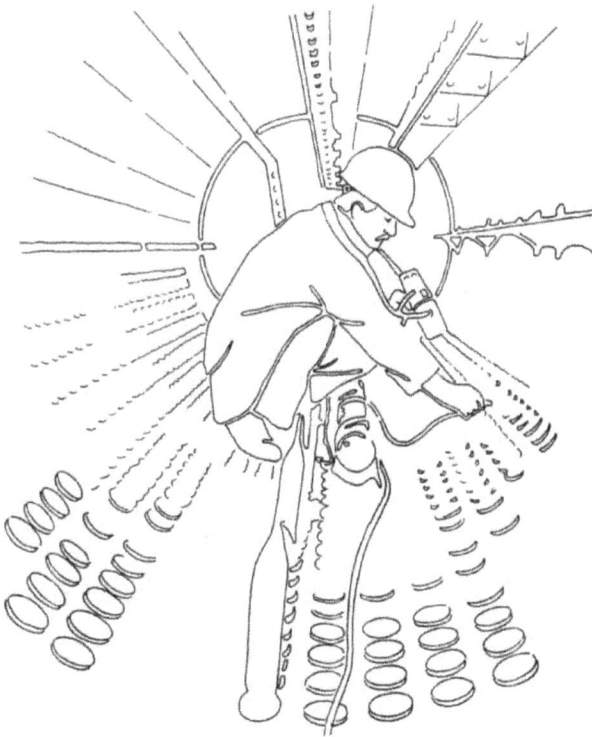

It is no accident," explained an article *in Materials Evaluation* magazine, "that the historical accounts of the history of boiler inspection are full of descriptions of accidents. One reason, no doubt, is the very human, if morbid, fascination that people have for those great natural and physical disasters that sweep humanity aside like dust.

"More importantly, such disasters turned people of science into people of action, promoting industrial reform and the adoption of codes for the benefit of both commerce and public safety. It was a long series of tragedies - of losses in property and lives - that gave the impetus to the practice and improvement of boiler inspection."

From the earliest days, nondestructive testing (NDT) in the form of *visual* examination was commonplace for the inspection of boilers and related components. Even before the destructive force of steam was fully understood, visual testing had become effective, certainly in determining how well equipment was being maintained, if not in locating inner defects in metals and fastenings.

A review of the very first ASME Boiler Code, developed in the years 1911 through 1914 and published in 1915, reveals that this Code - like its successors - provided guidelines for the inspection of boilers during construction. The "Rules for the Construction of Stationary Boilers and for Allowable Working Pressures" also advised inspection by the purchaser of equipment prior to installation. Hydrostatic testing was recommended for tubes, valves, and headers. For rolled bars, the standard stipulated a "chemical etch" test, applying a mixture of two parts water to one part hydrochloric acid. Tensile-drop and quench-bend tests were also advised by the purchaser as a means of testing broken flange and firebox specimens.

Among the duties of inspectors prior to World War I was the visual examination of malleable castings to determine that they were "true to pattern, free from blemishes, scale, or shrinkage cracks." (A variation of $1/_{16}$ inch per foot was acceptable.) The finish of flat bars had to be "smoothly rolled and free from slivers, depressions, seams, crop ends, and burns."

The inspector who paid proper attention to his business examined each and every part with care, to make certain that "the finished material shall be free from injurious defects and shall have a workmanlike finish."

Although the 1915 Code was sophisticated enough to stipulate such methods of testing boilers and other pressure vessels, it did not attempt to specify qualifications and procedures for the inspectors themselves. The Code Committee did not overlook this subject though, by any means. As was explained at the time, since many states had their own system for inspection and in some cases loosely knit organizations of professional inspectors, the

ASME Committee steered away from stumbling into "a mess of administrative and legal details."

The job of assuring uniformity in the procedures and qualifications of inspectors was thus left for a later date and in other hands. As has been mentioned earlier, this problem was effectively resolved in 1919 with the formation of the National Board of Boiler and Pressure Vessel Inspectors by a group of state enforcement officers and others concerned about the quality of inspections and personnel.

Even prior to the founding of the National Board, nondestructive methods other than visual examination had been developed, some with a fair degree of success. By the 1920s, when fusion welding came into limited use by manufacturers as a replacement for riveting, inspectors often used the nondestructive method referred to as "tapping" that has been mentioned earlier.

"When you speak of testing," explains Frank Williams, "you have to consider that you are talking about a many faceted subject. You have testing of materials, for which you are evaluating a whole series of characteristics that are vital in the performance of those materials under load. When those materials are joined, you have tests for welding - a number of different types. There is the surface test, the X-ray test, the sonic test, and in each one you are looking for certain internal characteristics or abnormalities. Then you come to visual tests, such as alignment tests, examinations for scratches and holes. And you finally come to a test for the strength of the whole structure. This is the most common and important, characterized by the hydrostatic test, putting the entire structure under internal pressure. That pressure is usually much higher - in most cases, 50 percent higher - than the pressure for which the equipment is designed. And there you get into a whole technical discussion as to why this test is good or bad."

During the 1920s, magnetic particle testing and induction testing, using direct current, were also tried, but with what was later described as "mixed results." Of greater value was radiographic testing (RT) perfected at the Watertown Arsenal by Horace Lester. This method was promising enough to be assigned to an ASME Code subcommittee for review and in 1931 the revised Boiler Code accepted welded vessels that had passed radiographic tests. It was at this time that magnetic particle testing finally came into its own, though largely as a complement to RT. As manufacturers continued to build larger and larger boilers that could be operated at high temperatures, the radiographic method had a tendency to miss certain types of surface cracks. Magnetic particle testing, however, was able to provide supplementary data and detect what RT had missed.

"RT techniques were further improved in the later 1930s," reported the editorial in *Materials Evaluation,* "when high-voltage RT was introduced. Certain defects, such as slag inclusions and lack of penetration became more evident. As a consequence, boilers became stronger and safer."

Ultrasonic testing (UT) began proving its worth in the testing of welds as early as 1945, and particularly by 1947 with the patenting of an angle-beam probe that was successfully tested on pressure vessels at the Combustion Engineering Company. Since that time, ultrasonic techniques have improved markedly, along with magnetic testing and radiography.

Even generations-old visual examination methods were still proving to be effective, enhanced by technological advances, such as borescopes and closed-circuit television that made it possible for inspectors to gain visual access to inner areas that they could not reach physically.

At the 55th General Meeting of the National Board of Boiler and Pressure Vessel Inspectors in Baltimore in May 1986, nondestructive testing (NDT) was briefly defined as "the industrial testing of material properties that does not impair the future usefulness of the test objects." Characteristic NDT methods cited were those that were visual, electromagnetic, radiographic, ultrasonic, and acoustic in nature. The organization most directly concerned with this technical field is the American Society for Nondestructive Testing (ASNT), originally founded in 1941 as the Society for Industrial Radiography. Its membership of some 10,000 professionals still includes many who entered the field of testing during and right after World War II, at a time when NDT was coming into its own and greatly influencing the rules and specifications being written into the Boiler and Pressure Vessel Code.

As Patrick Moore, an editor in this field, commented, "The first nondestructive test method was visual testing (VT), and the term VT refers, not to a cave man's inspection of his spearhead (although that is indeed VT), but rather to documented inspection of a product according to a particular procedure or specification designed to recognize material defects." The earliest such tests that were effective included the detection of corrosion on inserviceboilers, gross surface blemishes, and signs of poor workmanship or sloppy maintenance.

With every major boiler explosion or related disaster, efforts would be escalated to perfect testing methods that might be relied on to prevent future casualties and property damage. The effort paid off as the introduction of other, more effective, NDT methods began to make explosions rare. VT itself received a big boost when *assisted* visual testing was experimented with during the 1920s, resulting in an instrument referred to as the "borescope." Its predecessor - and the source of its name - was a device used to peer into the bores of military cannons to inspect them during World War I.

The terms *cystoscope, endoscope, telescope,* and *periscope* all relate to the same kind of visual-detection instruments used by doctors, navigators, and other specialists to enhance their natural vision during a sighting procedure. Adapted by engineers, these devices and scientific concepts could easily be used to see inside objects such as boilers, pipes, tanks, turbines, valves, and various closed vessels of many kinds. Ingenuity played a great part in perfecting the general class of borescopes through the addition of special lenses, mirrors, and lamps. Later, even computers would be hooked into a system to provide computer-enhanced images that could detect flaws that the naked eye could not.

Around 1930, a big step forward in NDT was taken with the introduction of radiographic testing, which made use of X-rays and similar instruments to seek out imperfections in metals and fittings that the eye could not spot, no matter how well "assisted" by instruments. Starting with the rudimentary examination of the seams of welded drums, this technology has developed from the use of a simple X-ray machine rated at 200 kilovolts to the application of 13 million volt linear accelerators, often supplemented by gamma-ray equipment, and capable of testing the most sophisticated nuclear reactor vessels.

The next significant NDT method to appear was electromagnetic testing (EMT), which was actually "a category of more than a dozen methods, including eddy current testing, permeance testing, magnetic flux leakage testing, conductivity testing, microwave/radar testing, and magnetic particle testing. Sixty years ago, the EMT methods were commonly called 'magnetic analysis.' Of these, eddy current testing is the most widely used and best known."

The theoretical groundwork for eddy current testing was laid in the 19th century when it became common knowledge that a magnet could be affected by a nearby electric current. As engineers sought to perfect better ways of using science to assist in the testing process, they experimented with ways to use magnetism and electric current - which could be precisely controlled and measured - to study physical objects. They found that under controlled conditions it was thus possible to make inferences about material properties on the basis of electrical signals.

Early electromagnetic devices made use of the telephone as the means of communicating results. Later, inspectors had at their disposal a range of other supplementary instruments to indicate the findings, such as oscilloscopes, cathode-ray tubes, meters, strip-chart recorders, magnetic tape, trip switches, and even bells and buzzers. During the 1930s, there was a considerable amount of eddy current research in progress in the academic world, particularly at Johns Hopkins University in Baltimore. However, "the true revolution in commercial eddy current testing" came in the 1950s when Friedrich Foerster began marketing his instruments outside his native Germany. Foerster, who was something of a legendary figure and in some disrepute in the United States because of his reputed role in inventing magnetic switches for explosive mines used by Germany during World War II, regained respect because of his non-military work. His theories and applications were described as "awe-inspiring contributions" to the ASNT's first edition of its *Nondestructive Testing Handbook*.

Radiographic Testing (1925-1950)

X-rays had been discovered in 1895 by Wilhelm Conrad Roentgen, though at first were considered to be of little use except for experimental work and amusement. Some 15 years later, however, the science became more respected when William R. Coolidge invented the X-ray tube and a fluoroscope not unlike the instrument used today to inspect luggage at airport departure gates. Systematic flaw detection, however, would have to wait until the 1920s, when Horace Lester at Watertown Arsenal would develop radiographic testing (RT) equipment and procedures for the inspection of materials as radiographically dense as 7.5 cm of steel.

Lester's work led to the acceptance by the ASME Code of welded pressure vessels for the highest pressures possible. It was this change in the Code that really gave impetus to the acceptance of radiographic testing by the industry. The Boiler Code Committee was said to have been influenced, too, by the successful operation of the first radiographic equipment ever installed in a boiler fabricator's shop, a significant event in its history that took place in 1930. The equipment was positioned on rollers, making it possible to record successive exposures along all parts of a weld.

Three further developments took place during the 1930s and 1940s that extended the acceptance and improved the reliability of radiographic testing: (1) portable X-ray generators were designed and installed to place the equipment in more widespread use; (2) gamma-radiography, generally using radium as the radiation source, was introduced commercially, and (3) high-voltage equipment with much greater penetrating power was perfected.

The introduction of portable equipment took RT into the great outdoors, X-raying things like pipelines and hydroelectric dams. Even so, the bulky components required for high-energy radiation testing continued to keep testing mainly in the laboratory, at least during the 1940s.

Ultrasonic Testing (1940-1960)

Credit for the first application of ultrasound to the testing of materials belongs to Sergei Sokolov (a Russian) and to O. Muehlhaeuser (a German).

Sokolov proposed the application in 1929, and Muehlhaeuser patented a device for ultrasonic testing (UT) in 1931. Sokolov was especially persistent in his UT research and visionary in the applications he saw for it. Credit for the first commercially feasible design, however, belongs to Floyd Firestone, a professor at the University of Michigan. His 1940 patent called for pulse echo flaw detection and saw its way into the marketplace some three years later under the name Supersonic Reflectoscope.

In the mid-1940s, another form of ultrasonic testing was introduced, referred to as "resonance" UT. The technique involved the creation of resonant vibrations in the test object, treating it much like a tuning fork, and was especially useful for the thickness testing of thin materials from one side and the detection of corrosion in piping.

The acceptance of ultrasonic testing by industry was inexplicably slow during the 1950s, despite these promising developments, including the first "angle-beam" instrumentation. By the end of the decade, the situation had changed and many firms in the ultrasonics testing field were described as "thriving."

Acoustic Emission Testing (1950-1970)

The phenomenon of "acoustic emission" had been observed even before the 19th century - the realization that sounds, though sometimes almost imperceptible, were emitted by materials undergoing changes in stress, chemistry, temperature, and other natural forces. The phenomenon was scientifically studied in a number of ways, including research in geophysical seismology undertaken in 1923.

The first research, though, that was to be of direct significance to the Code and the testing of boiler and pressure vessel components was a systematic investigation described by Josef Kaiser in Munich, Germany, in the 1950s. His work, and that of others in this field, led to the first practical and industrial application of acoustic emission in 1961 at the Areojet Strategic Propulsion Center in Sacramento, California. There, a research team used AE to test a pressure vessel that was designed for use with solid-propellant rocket motors.

"Before the pioneering work of Josef Kaiser on acoustic emission, it was already known that audible sound came from tin bars when bent," explained a profile of the scientist from the American Society for Nondestructive Testing. "As Kaiser prepared his Ph.D thesis at the Department of Mechanical Engineering and Electrotechniques at the Technical University of Munich, West Germany, he suggested that any crystalline solid would emit sound when put under mechanical load. He was convinced that either the intensity would be too low or the frequency too high to be recognized by the human ear. So Kaiser needed microphones sensitive enough to convert mechanical waves into electrical oscillations that could be amplified and displayed on an oscillograph. These were piezoelectric transducers."

During his early experiments, he placed transducers on the face of tensile test samples and recorded the emerging sounds as stress was steadily increased. These sounds, as it turned out, consisted of a variety of distinct pulses, rather than continuous waves. Tin samples produced sounds that were actually audible, whereas all of the other materials he tested - such as copper, lead, and carbon steels - emitted sound intensities that were several orders of magnitude below the limit of human audibility.

He found that the number of pulses in any given unit of time depended upon the velocity of stress application. Thus he came to the conclusion that sound was generated by many distinct processes of an irreversible, noncoherent nature. To relate these results to the stress-strain diagram of the tensile strength, he plotted the frequency and amplitude of each sample (as determined from an oscilloscope signal) in contrast to the stress that had been applied. Evaluation of the resulting diagrams yielded data that he could interpret accurately.

Kaiser thus reached the conclusion that sound was generated by shear and fracture processes within the crystallites themselves, as well as by the friction between crystallite surfaces. Since the individual crystallites possessed different orientations to the direction of applied tensile stress, the sound-generating processes would occur in sequence, and not simultaneously. In addition, tests further demonstrated that frictional heat caused the crystallites to expand, thus causing more acoustic emission. Even the application of very low stresses produced sound that could be detected and recorded.

Kaiser also experimented with the testing of metals that had been pre-strained and others that had been heat treated to determine how these actions affected the nature and intensity of the acoustic signals. Marked changes could be recorded that helped to determine the effect of stress and heat on the samples. These results represented a big jump forward in the development of materials testing. Until then, machinery and vessel components had to be examined by use of destructive methods to determine their residual lifetime or load capacity. From then on, the same could be achieved by the nondestructive method of acoustic emission testing.

Many of the advances in the field of examination and testing would never have been possible without the existence of the American Society for Nondestructive Testing, which describes itself as "a national technical society organized and operated exclusively for the purpose of advancing scientific, engineering, and technical knowledge in the field of nondestructive testing

127

through education, research, and the compilation and dissemination of information useful to the individual and beneficial to the public."

ASNT is a fine example of how the Code policies evolved and perfected a system of coordination with non-Code groups. As a matter of policy, the Code Committee had early on decided that it would not underwrite and conduct original research directly. While such participation could have resulted in important fact-finding results, the Committee did not want to place itself in the position of being faulted for sitting in judgment and perhaps being accused of not fairly evaluating research that it then would have sponsored and directed. This policy led to a series of very fine working relationships with other national organizations. The results speak for themselves because there are outstanding records documenting how all of these organizations have responded in a cooperative, timely, and positive manner to help the Code Committee keep its books and its requirements up to date.

ASNT publishes numerous reference works which are regularly studied by Code subcommittees working in this subject field. These cover such subjects as radiography, liquid penetrant tests, electromagnetic testing, ultrasonics, and acoustic emission. Nondestructive testing conferences and seminars are held regularly to review new refinements in testing, provide updates on research in progress, and examine the criteria for new types and applications of metals and other materials.

Handbooks and guides were compiled to provide specific information about duties and responsibilities. One such example covers the subject of components of boilers and pressure vessels that could be assembled and completed in the field. Initially, this kind of on-site work introduced some confusion about the matter of *responsibility*. Whenever the construction of a component that comes under the ASME Boiler and Pressure Vessel Code is completed in the field, it is quite apparent that *some* person or organization must take responsibility for the entire unit, including the issuance of data sheets on the parts fabricated in the shop and field and the application of Code stamping. Possible arrangements to satisfy the rules and specifications of the Code might include the following.

- The manufacturer assumes full responsibility, supplying his own personnel to complete the job.
- The manufacturer assumes full responsibility, completing the job with local labor under his supervision.
- The manufacturer assumes full responsibility, but engages an outside construction firm accredited by ASME to assemble the equipment, using the Code procedures specified by the manufacturer. With this

procedure, the manufacturer assumes the responsibility of making inspections to be sure the procedures are being properly followed.

- The installation may be made by a qualified assembler with the required ASME Certificate and Code symbol stamp, which indicates that his organization is familiar with the Code and that his operators are Code qualified. Such an assembler may assume full responsibility for the complete assembly in the field.

Any number of other arrangements, individually or collectively, are possible. But the important key is the issue of responsibility and the relationship to the Code. If a firm other than the shop fabricator or an authorized assembler is called onto the job, for instance, the extent of that association must be predetermined. This is especially important if, say, an outside contractor is hired to install external piping. The piping, often overlooked, lies within the jurisdiction of the Code and must be examined by the inspector who conducts the final tests before he can approve the master data sheets.

As nondestructive testing came more and more into the picture and more closely associated with Code publications and procedures, the different techniques became better known for certain applications than for others. Here are some typical applications for the major techniques of nondestructive testing.

- *Radiographic inspection (X-ray)* - used for the examination of the internal soundness of welds, castings, forgings, and plate material
- *Radiographic inspection (gamma ray)* -used for much the same purposes as X-rays, but more suitable for heavy thicknesses or where portable equipment is an advantage
- *Magnetic particle inspection* - locates surface and subsurface defects that are not too deep but that tend to be overlooked by radiographic testing; especially useful for the inspection of nozzle and manhole welds, weld repairs, and the detection of laminations at plate edges
- *Penetrant (dye) inspection* - the most widespread application is for surface defects, nonmagnetic materials, and the inspection of nozzle and manhole welds when radiography is difficult
- *Penetrant (fluorescent) inspection* -useful for locating defects that run through to the surface; especially applicable for nonmagnetic materials, rough surfaces, and the detection of porosity that would otherwise be difficult to locate
- *Ultrasonic inspection* -mainly used for the detection of defects in welds and plate material and for determining the thickness of plate;

129

advantages are that it can access from one side of a part being tested and that it reveals small root cracks and defects not picked up by radiographic film, especially in thick-walled vessels

CHAPTER 13

KEEPING ABREAST OF TECHNOLOGY

Breakthroughs and Milestones

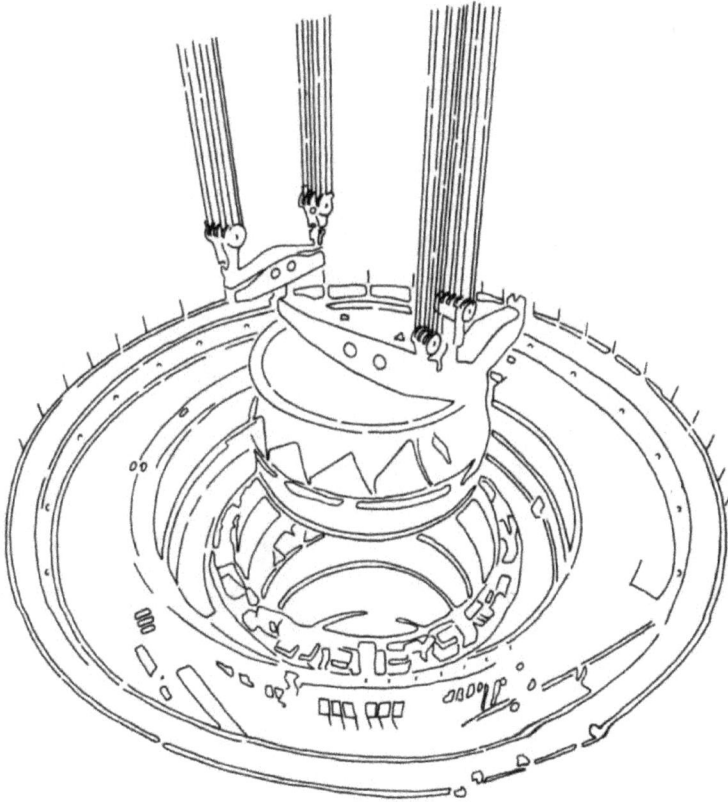

Throughout the years, the greatest challenge to the ASME Code Committee and the many subcommittees was to keep abreast of ever-advancing technologies so that standards and specifications would reflect the current state of the art in all subject fields covered. Accomplishing this major objective required not only additional committee members from one year to the next, but also the inclusion of specialists who were fully experienced in new fields of research and development pertinent to the subject areas covered by the Code.

The numbers were to increase dramatically following the hiatus years of World War II when many research projects were temporarily frozen, and particularly with the arrival of the age of nuclear power. By the mid-1970s, according to an article in *Iron Age,* the Code was being written and kept up to date by "a volunteer army of some 750 engineers, practically all of whom hold down full-time jobs in the private sector. No group is more familiar with the technologies embodied in the Code."

The article also emphasized a significant — and, indeed, rather astonishing — fact: To the best of the publisher's knowledge and after some exhaustive editorial research, no one had been killed by an explosion blamed on poor design or faulty construction in any jurisdiction where the Code had become law. Whatever fatalities had occurred in recent years could be traced to improper operations of boilers or pressure vessels. And there was no way that any code, however stringent, could obliterate human error.

Facts like these spoke well for the diligence and expertise of the many code committees that had come into being during the 1930s and 1940s when widespread and complex technological advances were burgeoning all across America and in lands abroad that were affected by the Code.

By this time, the representatives serving on the Code committees had used the organization's years of experience wisely and well. The principal power rested, as it had in the past, with the Main Committee, which was charged with the responsibility of giving approval to additions and amendments proposed by subcommittees. In order to tackle this mandate fairly and squarely, it was vital that the Main Committee be as free as possible from any hint of self-interest or conflict of interest. This concept was described accurately by Leonard P. Zick, now retired, in an account of his experiences as a former chairman of the Main Committee.

"We tried for, and usually had, balanced representation on the principal committees," he explained, pointing out that his group had been comprised of six fabricators, four makers of steel and other materials, seven industry owners, six from the ranks of the federal government and other jurisdictions, and six drawn from the fields of education, consulting, research, and engineering. These representatives across the board were then meeting, not once or twice annually, but six times a year to make certain that they were not letting any grass grow under their feet.

"It's more than a code," said Zick. "The related groups make up a safety system. Our main objective is to provide requirements for new construction of pressure-retaining units that, when followed, will provide safety to those who use them and those who might be affected by their use. And, since the use of the code item might be for any type of process or for any discipline involving energy, the committee's activities do not play one against the other. We want *all* code items to be safe. Period!"

During the 1940s, and continuing strongly thereafter, the Code came under attack by critics as leaning strongly in the interests of "big business." It was a subject that recurred, mainly because there were, and always would be, incidents in which small manufacturers or users would complain that the rules and specifications, by their demanding nature, tended to generate higher prices or inflict increased expenses. While these could be absorbed by large corporations, it was inferred, they played havoc with small companies that had limited budgets.

Zick explained to would-be detractors that "The ASME Code is the only code requiring independent, third-party inspection. This requirement hardly represents the selfish interests of big business. Code stamps have been taken away from some companies, large and small, because they were not adhering to the Code."

The third party mentioned was the National Board of Boiler and Pressure Vessel Inspectors. The Board, now headquartered in Columbus, Ohio, had earned growing respect since the time of its founding in 1919. Because its membership was comprised of chief inspectors from various jurisdictions in the United States and Canada, it acted as a sounding board that would instantly and clearly echo evidences of self-interest or conflict of interest. The National Board had also set very high and objective standards to insure that its work would complement the Code Committee's built-in system of checks and balances. It was mandatory that, in order to perform Code inspection, individuals had to earn commissions by passing an examination that was prepared and marked by the Board. They were also required to have had at least one year of related experience if they held a degree in mechanical engineering or three years of related experience if they had no such degree. Many of these inspectors were employed by insurance companies, whose basic objectives were to reduce accidents and maintain high degrees of safety.

"The voice of inspection plays an important role throughout the labyrinthine structure of code committees," said the *Iron Age* editorial. "There is also the conference committee, which is made up of the chief inspectors from the various jurisdictions, each of which has a vote on the first action of the Main Committee."

Because of all these factors, *impartiality* had become an important byword for the ASME Code Committee.

From time to time, there have been movements underfoot and motions on the floors of the United States Senate and House to let the government take over, or at least regulate, all development of codes and standards. These actions come about because, as one ASME engineer expressed it, "our toughest job is not how to build safe plants and equipment, but how to convince the public that we know what we are doing. We all shudder to reflect on how inept the government would prove to be if it tried to inflict bureaucratic procedures on an industry that already must focus its every waking moment on the immense technological changes taking place."

It had been evident to almost everyone concerned that the passage of a government takeover bill would be disastrous. Code matters would be relegated to civil servants, many of them — perhaps *most* of them — with no knowledge of the inherent technology. Even if the government were to hire experienced engineers and specialists from industry to administer the Code, these people would soon lose the valuable day-to-day contact with industry that is so vital, and in many cases would not even be aware of current problems that needed to be considered in the drafting of rules and specifications.

"An affront to personal liberty . . . ."

"Un-American..."

"A dangerous form of socialism. . . ."

"The work of the devil. . . ."

Throughout the years, such phrases have become familiar to the ears of Code committees who find their work attacked by critics. The above might have been said in 1930 or 1940, or later. In point of fact, they were the prototypes establishing the pattern, for they were quoted back in 1912 — and by some industry leaders at that, whose target was the pioneering group of ASME members then trying to establish the first Boiler Code.

The position of ASME whenever such issues surfaced in recent years has been to cite the American National Standards Institute as the coordinator of codes and standards, rather than the government. That was the only rational and practical way in which *voluntary* standards could be maintained. The Boiler and Pressure Vessel Code has been an American National Standard for the past 16 years. The Boiler and Pressure Vessel Committee over most of its history proudly operated independently of ANSI. It was as independent toward ANSI as it is now toward government.

In the area of design, one of the most controversial subjects had long been whether to design to *tensile strength* (the resistance of a material to longitudinal stress, measured by the minimum amount of longitudinal stress required to rupture the material) or to *yield strength* (the stress necessary to produce a given inelastic strain in a material). After World War II, metallurgical laboratories were researching and developing an innovative series of new high-strength, low-alloy steels that looked particularly promising for pressure vessel applications. Many such steels were already being used in European countries, where designs were based on yield strength.

As the ASME Code evolved, it specified that designs were to be based on both tensile and yield strengths. "It's a poor argument to base your considerations on yield strength alone," say members in support of this viewpoint. "More properties are involved than just yield strength. If fatigue is a consideration, tensile strength is more representative than is the yield. Even the European proponents of designing to yield will use ultimate tensile or some other method to penalize material with high yield-to-ultimate ratios."

Another issue was timeliness and the occasional complaint that getting approval on Code changes relating to new technology was sometimes painfully slow. Yet the Committee procedures had been carefully studied and developed. Changes in the Code followed two routes. One was the Code Case, of which the Committee was then processing about 50 during the course of each calendar year. These represented changes that could be implemented within several months and which were permissive. The other was the Code Addenda, which was published biannually and which moved much more slowly from the date of origin to the time of approval, ranging from about six months to two years.

As was the case with new material, ASME required a specification from the American Society for Testing and Materials before the material could be included in the Code. This was traditionally accomplished through an addendum. A Code Case could cover a new material until it obtained an ASTM specification, after which it could be adopted into Section II of the Code. The actual material Code Case had to be supported by the required material and welding test data, and evidence that the material would be used in a Code application.

From the standpoint of private companies, the Code had already established itself as the basis on which to aim for quality and reliability, even in cases where the Code was not mandatory. "We live by the Code," said Paul M. Brister, before his retirement from Babcock & Wilcox, which had played such a key role in the original creation of the ASME Boiler Code, "for our boiler, pressure vessel and nuclear steam supply business. Even when code construction is not a specification requirement, our company standards require code construction as a minimum. From a materials standpoint, all materials used in equipment for pressure containment must meet ASME or ASTM specifications as a minimum.

"As far as the development and use of new materials in our products are concerned, an evaluation must be made on the length of time it will take to gain code acceptance. The code committees require test data on the properties of new materials from at least three commercial heats. When elevated temperature properties, such as creep or stress rupture strength, are required, it may take from two to five years to obtain the data and receive code acceptance."

It is not unusual for a manufacturer or designer to pioneer in the development of materials or products long before they are written into the Code — or even considered for such approval. A characteristic case was that of a fabricator who was interested in a promising new welding technique, so much so that he had invested a substantial sum in research and development. The fabricator foresaw a time when the technique would improve his products and would be a valuable asset in Code construction work. It was obvious from the start that a Code Case would eventually have to be submitted to the Code Committee for review, approval, and publication. First, however, a procedure qualification had to be completed — one that would satisfy the expectation of the Code administrators that the quality of the weld metal and the heat affected zone would be equal to that of the base metal.

One of the major trends over the years has certainly been the establishment of communications and the development of cooperation and coordination with other organizations concerned with the safety, efficiency, and reliability of boilers, pressure vessels, and related components and materials of all kinds. A good example is the relationship of the ASME

Code committees to the Pressure Vessel Research Committee of the Welding Research Council. So many significant developments have taken place that it is difficult to single out those that could be considered milestones. However, when asked to describe its major contributions to the Code a little more than a decade ago, the PVRC mentioned two areas in particular: reinforced openings and fracture mechanics. It was pointed out that the weak point in any pressure vessel was its reinforced opening. Much of the PVRC's research in recent years had been directed at reducing the concentrations of stress around these openings through improvements in design. A great deal of the contemporary research was being undertaken in the field of fracture mechanics and plans were being made to address more attention toward the behavior of steels in the creep range. In addition, the Research Committee had already initiated a major international program in the nondestructive evaluation of welds that involved participating groups from the United Kingdom, West

Germany, the Netherlands, France, Sweden, Belgium, Italy, Japan, and Canada.

A considerable body of research had been undertaken, too, on gaskets and postweld treatment. The basic gasket technology for flanged joints had been criticized as being behind the times and, indeed, was at this juncture still based largely on a paper that had been published in 1937. It was anticipated that a reassessment of gasket technology could very well result in entirely new ways of designing these, and related, products.

As for postweld heat treatment, there were indications that mandatory stipulations in the Code might be outmoded and that certain thicknesses of metal, for example, might not require heat treatment at all under certain circumstances. The reverse might also be true in situations where treatment was *not* mandatory but might produce more effective results. In any case, since the procedure was both time-consuming and expensive, it was the committee's consensus that a "more realistic" approach to postweld heat treatment was in order.

The ASME Boiler and Pressure Vessel Code Committee also worked very closely with the Metal Properties Council. The data accumulated by the MPC were often compiled only after long periods of study and investigation by broad gatherings of metallurgists and others in this field who brought various problems to the Council's attention. The data thus compiled achieved great credibility and acceptance and were highly influential in charting the future course of the Code.

A characteristic example of the Council's trend of work was the compilation of a considerable amount of property data on chrome-moly steels, which had been used to establish allowable Code stresses. Other major studies that were, or would be, significant involved continuing studies in the field of nuclear energy, materials under consideration for coal gasification plants, and the effects of varying temperatures and environments on the properties of metals.

Whether working independently or with other major organizations like the PVRC and the MPC, one forecast was certain. Although 11 sections to the Code had already been promulgated, the most recent addition having been Division 2 of Section III covering the permissible use of concrete in the design and construction of pressure vessels, additional sections would be forthcoming as soon as advancing technology dictated their need.

During the 1930s, keeping abreast of new technology had largely been a matter of focusing on the steadily increasing demands for components that could function safely at higher and higher temperatures and when subjected to greater pressures. Allied with this was continuing research to explore new alloys and the practicality of joining steel plates with ductile welds that were stronger than the plates joined, thereby eliminating the necessity of increasing the thickness of vessels at the joints.

"In the building of vessels for high pressure and high temperature," reported an article in *Mechanical Engineering* in March 1930, "it is becoming increasingly necessary to carefully reconsider the questions of design and construction from the point of safety because of two very important reasons. First, because the engineers for the steam-power, oil-refining, and chemical industries desire to make use of the economies rendered possible by going to higher and still higher pressures and temperatures. Second, because the problems associated with the safety of the equipment under those increasingly severe service conditions are becoming more and more difficult as the temperatures and pressures are increased."

It was evident during this era that efficiency and economy of operation, which had long been almost secondary considerations, were now factors of prime importance. For those working on or administering the Code, the situation was almost the reverse of what it would be in later years. Instead of bemoaning the onrush of technological change, as would be the case after World War II, mechanical engineers in this field were seeking out and encouraging breakthroughs that would permit the Code to permit greater extremes of pressure and temperature. There was also that strong trend toward larger and larger boilers and pressure vessels, which we observed earlier, and which had improved the economy and efficiency of many commercial operations.

The four most important points in considering vessels for severe service were, in order, as follows.

- *Design,* which, if improperly or carelessly conceived, could result in the utilization of less than half the strength that the materials could produce. The critical features of design were considered to be the shape of heads, proper reinforcing, and the efficiency of the joints.
- *Materials,* whose strength and capabilities must be totally understood in order that full advantage might be taken of the types and grades available and their suitability for the operating environment and conditions that would prevail during service.
- *Fabrication,* in the best manner possible that would bring the selected materials into conformity with the design. It was emphasized that excellent materials and outstanding design would be negated if the fabrication did not live up to expectations.
- *Testing,* to determine the quality and capabilities of the finished vessel, using proven methods that in the laboratory could simulate on-

site operation under extremes of pressure and temperature without risk to property or personnel, should a failure occur.

Many developments were also taking place with regard to turbines of all kinds and types, for a wide variety of uses. A review of the design and performance of large steam turbines at the beginning of the 1940s indicated that the trend during the previous two decades had been steadily toward the design and construction of units, including attached generators, with ever-higher capacities to meet the growing demands for more power and to permit reduction of the plant investment per unit of output. This trend was complemented by the continuous modification of design, manufacturing systems, and materials in order to increase the reliability of operation and to reduce outage time, which was a constant concern.

The utilization of higher initial steam temperatures and pressures, coupled with improved heat cycles (such as regenerative feedwater heating, preheating, and resuperheating to decrease fuel consumption) was also a key factor. Finally, the records show increasing attention over the years to the design of turbines to utilize ever-increasing proportions of the available energy in steam cycles being used.

"The major endeavor," explained a contemporary report, "has been to improve these factors through the use of new knowledge, new materials, and new tools and processes of manufacturing." The study, concluded in 1941 by the Turbine Engineering Department of General Electric, evaluated the tests on more than 100 steam turbines under station-operating conditions and concluded that great progress had been achieved during the previous 20 years. This was especially valid in the matter of operating efficiencies, economy, and the ability to function for long periods of time with no outages, despite being subjected to greater pressures and temperatures.

This was the period, prior to World War II, when turbines were being designed for some of the nation's great hydroelectric power plants. The Grand Coulee power plant was a case in point, the largest hydroelectric power plant in existence, containing 18 150,000-horsepower generating units and three 14,000-horsepower station-service units for a total capacity of 2,742,000 horsepower. Grand Coulee provided for, among other units, 12 150,000-horsepower turbines capable of driving 60-cycle main-unit generators of 108,000 kva each, for this historic Columbia Basin project.

The planning and eventual completion of Grand Coulee and other major hydroelectric plants in the United States served to focus the nation's attention on the increasing demands for electricity and power, as well as to understand and respect more fully the engineering profession and the many challenges it faced.

By the end of the 1940s, the gas turbine had come into its own and was now the subject of widespread research and development. "The advent of the gas turbine has put a new emphasis on the development and testing of high-temperature materials," wrote A.D. Hughes, a professor of mechanical engineering at Oregon State College. "Means must be developed to predict the behavior of such materials under actual service conditions .... Static tests, isolating one or two of the variables, such as rupture strength or resistance to oxidation at high temperatures, provide important data, and these cannot be neglected. Nevertheless, in an actual machine, service life of a material is usually affected by several factors at the same time. On this basis, it is felt that essentially full scale tests under service conditions will give data more applicable to the future design of longer-life gas turbines."

Dr. Hughes pointed out one of the key factors that had brought commercial gas turbines to this state-of-the-art condition: the splendid performance of gas-turbine-driven superchargers used in American bombers and fighter planes during World War II. He looked forward to the time when gas turbine power plants could compete with the efficiencies that had been demonstrated in recent years in stationary steam or diesel plants. Much of their success would depend upon the capability of the experimental units to prove themselves during tests.

The history of gas turbine development plainly showed that the earlier models were "doomed to failure" because much too large a percentage of the power produced by any given turbine was wasted in trying to drive the inefficient compressors that were then available, leaving few or none to do the work. It was not until the start of the 1940s that this problem began to resolve itself, with the design and manufacture of more effective compressors and with the availability of metals that would withstand operation at 1,000°F for extended periods of time. Although many private-industry research projects had to be placed on the back burner during the war, the development of the gas turbine did benefit from military experiments and research, particularly by the U.S. Navy and the Air Corps. A number of these were concerned with alloys that were created to survive sudden pressures under duress. Outstanding, too, were alloys that had a tendency to retain their ductility when exposed to elevated temperatures, which was almost mandatory in minimizing the effect of repetitive thermal stresses that occurred upon the starting and stopping of large turbines.

Early in 1946, secrecy restrictions were lifted on the so-called "super" heat-resisting alloys perfected for military applications, releasing them for use by private industry. With this new state of the art available, mechanical engineers were able to take a quantum leap forward in a number of areas that related to energy and power. Such information also benefitted the administrators of the

ASME Code since it placed in their hands valuable data that could be written into the rules and specifications for boilers, pressure vessels, and related components.

Another major trend in the development of turbines and power-generating equipment of all kinds was the use of various supervisory instruments and controls. This concept became more and more vital because of the increasing move to centralize controls at points remote from the power-generating equipment itself. It was also an outgrowth of "the growing trend of power companies to quick-start their machines," according to a research report from General Electric in 1952.

By this time, sophisticated equipment had been perfected and made available that could, by remote control, monitor turbine performance and conditions in terms of shell and differential expansion, speed and camshaft position, bearing vibration, and shaft eccentricity. The value of the instrument depended, of course, upon the proper interpretation of the data transmitted.

"The trend in the operation of power stations in recent years," explained the GE report, "has been to install power-station controls in a central control room at a point remote from the turbine. Thus it follows that the supervision that is so important while starting the machines, and during their running period, is substantially reduced. Consequently, it is imperative that adequate instrumentation be provided to indicate to the operators the condition of the unit. In conjunction with this is the growing tendency for power companies to put their equipment on stream quickly and sometimes abruptly. This stems from the large drop-off load at night, in which the less efficient units are shut down and then quick-started in the morning to satisfy the rapid increase in load demand."

Recognizing these trends, GE and other companies in the industry realized that more sophisticated and more comprehensive instrumentation was becoming necessary in order to keep track of the rapidly changing situations in areas remote from the centers of control. The most vital requirement and also the most difficult problem that had to be surmounted in the perfection of these instruments — such as strip-chart recorders, electronic power units, and suitable detectors — was to design them so that they had accuracy and reliability factors comparable to that of the turbines they were supposed to monitor. A number of actions made this possible, including the design of instruments that were conservative and could be upgraded as the art of electronics advanced; attached test circuits to check out the instrumentation periodically; and built-in calibration checks by which the overall calibration could be steadily maintained.

Contradictory though it may have seemed, one of the considerations of engineers desirous of keeping abreast of advancing technology had to do with the burning of coal. This was a subject of particular concern to the administrators of the Boiler and Pressure Vessel Code since coal was coming into its own again as a fuel for boiler installations. Among the factors motivating the resurgence of coal were wartime lessons in fuel burning; the use of pulverized coal for generating steam for turbines; coal gasification research; and the dramatic improvement of equipment for processing, feeding, and burning coal of numerous grades and in several forms.

"The design of postwar power plants reflects improved technology, particularly with respect to fuel utilization," reported an ASME paper on "The Improved Application of Coal Burning Equipment" in 1947. "The restricting effect of the war and the depression which preceded it on the number of new installations had its fortunate side. The war taught many lessons regarding fuels and their utilization and the net result is that the installations being made today are better than had they been made prewar. This is particularly true with respect to fuel flexibility, since the modern installation is much more flexible fuelwise than was its predecessor."

The enormity of coal reserves and the fact that bituminous coal was the basic fuel for steam generation led to a trend toward providing better grades of coal on the one hand and designing more advanced equipment for utilizing it on the other. New pulverizers and improved automatic stokers were among the major results of this development. Keeping pace were research projects to provide more effective and more economical design for coal-burning furnaces. These studies, in turn, directly affected boiler design, to take advantage of the available fuels, in this case coal and derivatives of coal. And all of these combined developments, of course, had a direct effect upon the scope and content of specific areas of the ASME Code.

The pulverized-coal-burning open-cycle gas turbine was an example of the kind of development that was taking place during this trend in the history of energy.

Coal gasification was another state-of-the-art development at the beginning of the 1950s. According to a 1952 paper from the Carnegie Institute of Technology, "Intensive research and development work during the past few years have demonstrated the feasibility of producing carbon monoxide and hydrogen by the continuous and complete gasification of pulverized coal with oxygen and steam. This has led to considerable interest in the thermodynamics of such systems."

In the course of experimental gasification work at the Bureau of Mines in Pittsburgh, extensive material and heat balances were made to compare the results with the theoretical values and to provide guidelines for using this kind of fuel for optimum performance. These data were then applied later to experimental designs for related equipment.

Over the years, the Code Committee was continuously challenged by the increasing number of inquiries it received about interpretations of the rules, as well as by suggestions and recommendations for additions and revisions. This was only natural, in light of the fact that each successive Code Edition was larger and more complex than the previous one, a reflection of technological advances and the development of new products and appurtenances. As users were given more and more choices of products and materials, the work of the main committee and the subcommittees escalated accordingly, year after year.

At the end of the 1930s, positive steps were taken to try to find a solution to this problem, an effort that had been made several times in the past, yet without notable success. The catalyst in this instance was the National Bureau of Casualty and Surety Underwriters, which had sponsored an Engineering Conference in 1938 to discuss matters of mutual interest to insurers and engineers. At a meeting of the Boiler Code Committee that year, a spokesman for the NBCSU, J. P. H. de Windt, read minutes from the Conference opposing the great numbers of revisions and interpretations to the Boiler and Pressure Vessel Code. It was becoming more and more difficult, the statement asserted, to keep abreast of the changes without constantly doing an exorbitant amount of homework. Moreover, in many instances when accidents happened or equipment failed, it was difficult to determine who was at fault when insurance claims were initiated or there was a legal issue because of the complexity of rules and specifications. In a more routine way, the on-the-job performance of insurance company inspectors was suffering because they could not keep up with the revisions. One faction within the Bureau was convinced that far too many changes in the Code were prompted by *economic* considerations, to solve problems of manufacturing costs rather than to improve what were fundamental safety measures.

The Bureau prepared an official memorandum of opposition for presentation to the ASME Code Committee, recommending that all interpretations and revisions in the future be carefully analyzed from a viewpoint of safety. In cases where production or installation costs or other economic factors were the basic problems and the issue of safety was nonexistent or at best insignificant, no revisions should be authorized. It was pointed out during the discussions that the National Bureau of Casualty and Surety Underwriters was not alone in trying to cope with the frustrations of too many interpretations and rules. Members of The National Board of Boiler and Pressure Vessel Inspectors were equally distressed, particularly those who were engaged in field work and the day-to-day checking of equipment against the Code. Administrators of the National Board were worried in a different way, having already heard negative comments from representatives of some jurisdictions that the Code was becoming too ponderous. It would take only a few dissatisfied states and municipalities to upset the balance coast to coast, since their refusal to accept certain changes would then be out of synchronization with the rest.

The outcome was not as conclusive as many of the people involved might have hoped. The Code Committee justified some of the types of revisions cited as examples of overkill on the assertion that they provided improvements in design and on-site performance, which were certainly Code objectives. However, the Committee members were in general agreement that careful attention should be paid to all suggested revisions and that, indeed, if the reasons were mainly to reduce manufacturing costs, no revisions should be made.

CHAPTER 14

THE CODE GOES TO WAR

Crisis Situations Test Concepts and Usages

With the arrival of the Second World War came what one engineering historian, Ernest L. Robinson, called "the urge to accomplish the impossible." As he related, "American warships, propelled by higher temperature boilers and turbines, had cruising ranges 50 percent higher than their British friends, who had to refuel three times to our twice." In the air, American airplane engines equipped with superchargers had an extra increment of power and speed that gave them a definite edge in combat. "These superchargers had a wheel and buckets of super-strength high-temperature alloy capable of running red hot for 1000 hours," said Robinson. "The supercharger was a true gas turbine although it lived with and became part of a piston engine."

The gas turbine on its own saw its first practical application during the war, whose crucial and immediate demands actually motivated the supreme effort that led to the creation of super-strength, high-temperature alloys necessary to build the engine.

British engineers had started working on improvements of the jet engine, which was revolutionary because it performed poorly at low speed and reached its peak performance only at high speeds. Until the war put a demanding premium on high speed, there were plenty of reasons to do nothing about it.

Then came a memorable day when a coded telegram from Buffalo announced, "Baby safely delivered — mother and child doing well." This was code language to report that the first jet airplane engine had been delivered to the manufacturer of the plane. The phenomenal success of the jet engine, both for military uses and later for civilian transport, has long since become a matter of history.

Even long before World War II, production for the "Arsenal of Democracy" had greatly accelerated the expansion of American industry. The pattern of World War I reasserted itself as steam-generating equipment was hastily called upon to power the machines of war. In the marine field, power plant developments of the restrictive pre-war years paid an important and unexpected dividend. Designs that hadbeen researched and evolved for boilers were being built as the U.S. Navy program — beginning in 1933 — gradually gathered momentum.

Power was needed for what was to be the greatest naval and cargo fleet ever assembled. The industry, with low production at the end of the 1930s, suddenly had to go into high gear. It was estimated that, from the time of Pearl Harbor until the end of the war, about 5,500 major combat and supplementary merchant vessels were constructed in America that were powered by steam. They included aircraft carriers of all classes, battleships, cruisers, and hundreds of cargo vessels and destroyers. One indication of the power needs of some of these ships was reflected in an official Navy release describing the supercarrier *Coral Sea*. Its state-of-the-art boilers were capable of generating enough power to satisfy all of the requirements of a city with a population of one million people!

During the course of the war, the industry devoted many of its efforts and committed many of its facilities and manpower to projects that would help public utilities and priority industries improve the efficiency of furnaces and boiler operations and thus strengthen production lines. Fuel economy was especially critical when it was considered that such a heavy proportion had to be diverted to military operations. A ship like the aforementioned *Coral Sea,* for example, had to carry in her bunkers a supply of fuel oil that in peacetime would have heated some 3,000 average-sized homes for a year.

This dedication to the war effort, however, was destined to pay off in the immediate postwar period. One dramatic evidence of this was a new type of furnace, which was designed to solve two basic objectives desired by power engineers. The first was to use low-quality, high-ash fuels for the generation of steam without causing a build-up of undesirable residue. The second was to contain the coal ash within the furnace in the form of a molten slag which could be more readily controlled. Accomplishing these objectives was obviously going to be even more desirable when air pollution would become a public issue and power companies would come increasingly under attack by environmentalists for releasing harmful substances that imperiled the environment.

The answer to both of these questions lay in the perfection, just before the end of the war, of a furnace that was designed to burn crushed coal. When mixed with primary air and then "scrubbed" with secondary air, the coal produced such rapid combustion and such great heat that the ash melted instead of leaving the furnace with escaping gases.

The performance of this new furnace made possible improved designs in boilers, more advantageous operating conditions, minimal air pollution, and substantial reductions in both capital and operating expenses.

On the technological front, these advances were quite memorable and often dramatic, giving engineers in many fields a great sense of achievement. This elation was not experienced on all fronts, however. An historical account from the Hartford Steam Boiler Company summed up the situation. "With the entry of the United States into World War II, industrial production underwent a sudden acceleration. The company had an important contribution to make in seeing that the country's power equipment and production machinery were properly maintained for maximum production as well as for safe operation. With factories running around the clock, failures were inevitable. HSB was actually operating with a reduced engineering staff. It was also contributing time and manpower to the Federal Government, as it had done during the previous war. The maintenance problems, the breakdowns, the need for replacement parts surpassed the capacity of HSB to keep up with them, and the company suffered heavy losses. From this experience, HSB learned ironically that peak levels of industrial activity are not always good for business."

As many a major insurance company discovered to its chagrin, the war effected a radical change in the attitudes and policies of manufacturers. Because they were operating in many cases around the clock and completing military contracts, they openly balked at shutting equipment down long enough for the kinds of Code-related inspections that had been routine before Pearl Harbor. In addition, the "cost plus" type of emergency contracts that manufacturers held with the Federal Government led management to consider much of the company's equipment to be *expendable.* Such situations proved very costly to the insuring companies.

Turning for help to the ASME Code Committee and other bodies involved with establishing and supervising codes and standards brought little or no relief. In fact, the Boiler and Pressure Vessel Code Committee was itself the target of great pressure from manufacturers to modify specifications, waive certain rules, and permit the use of non-Code materials as substitutes. Furthermore, the insurers emphasized that not all of the difficulties could be attributed directly to the war. As one expressed it, "The boiler and machinery industry is in a confused and confusing situation, many aspects of which are of long standing."

One of the areas of specific and ongoing concern to the Code Committee, and which was greatly affected by the tides of war, was the matter of materials and materials specifications. Prior to 1940, the Subcommittee on Materials covered cases and specifications for all materials, ferrous and nonferrous, cast and wrought, as well as stress allowances. By the early 1940s, the nonferrous materials were split off under a separate group. This separation represented a major change in the Committee structure and recognition of the fact that, as the chemical industry developed and corrosion control became a part of so many processes, the Committee had to have both depth and breadth of competence in nonferrous areas to fulfill its obligations.

During World War II, the Main Committee decided to raise the allowable stresses on all materials (below the creep range) from a denominator of *five* to *four.* The reason for this decision was not that the materials had become stronger but that, with improved designs and records of excellent performance, the materials could safely be subjected to higher stresses, thereby helping to conserve materials. This was especially important at a time when conservation was critical and when some of the rarer metals were those that were most in demand for sophisticated controls and devices.

From 1940 to 1970, the number of materials specifications approved for Code use grew rapidly. Around 1940, Section II of the Code covered some 55 specifications. By 1970, this number was totaling about 200. As a result, Section II expanded into the largest section in the entire Code, becoming so unwieldy that, shortly after World War II, it was necessary to divide the book into three sections: Ferrous, Nonferrous, and Welding Rod.

After the War, too, Section III and Section VIII, Division 2, were created, each with a safety factor of three.

Another subject that was dealt with was brittle fracture, which was largely associated with welding. During the 1930s, the assembly of steel plate during construction by welding rather than the conventional method of riveting came into common usage. It was not altogether successful at first, resulting in restraints and defects that had been hitherto unknown. With the arrival of World War II and the rapid escalation of plans to build a sizable merchant fleet, the situation became compounded. The hulls and decks of cargo ships built during the early part of the war began cracking and failing by the dozens.

"Many of the failures were brittle fractures that started at corners of large hatch openings," explained Leonard Zick. "These and other failures led to Code revisions on fabrication and the examination of nozzles."

Faulty design, inexperienced workmanship, and haste all contributed to the problems. But the strains induced by welding during construction and the lack of flexibility of the welded hulls were probably the major causes of trouble. As a Committee member expressed it, "Pandora's box had been opened and the whole brittle-fracture problem descended upon our heads." As a result, great efforts were made to find solutions, which came about largely in the form of new steels with lower transition temperatures, improved ship design, and new theories about the origins of brittle fracture.

As has been previously covered, the 1930s had seen substantial increases in the steam temperatures in boilers, along with more formidable pressures. It was eventually realized by the industry that these changes required stronger and more creep-resistant steels, which were already being produced by the time the nation was girding for war. Molybdenum steels became popular for steam piping in central generating stations and for other uses, since it had proven to be highly resistant to creep. However, it was then found to be susceptible to graphitization, particularly in the vicinity of welded joints.

In the early 1940s, one of the most clear-cut examples was revealed when a major power failure occurred at the Springdale, Pennsylvania, generating plant of the West Penn Power Company. A large heavy-wall steam pipe, 11" in diameter, completely separated adjacent to a welded joint. Fortunately, instant emergency measures cut off the steam before there were any catastrophic failures or injuries to personnel on duty. Microscopic examination later

152

revealed that the carbon in the steel near the weld had migrated to a narrow band, resulting in a flawed strip that had almost no strength. Graphitization was also detected in carbon-molybdenum steel piping in more than 20 other power stations. Although the cause of this kind of graphitization was never precisely determined, it was found that the addition of 0.5 percent of chromium to the steel and some modifications in the steel manufacturing procedures would eliminate the problem completely.

The Boiler and Pressure Vessel Code Committee was known to have long been traditionally "allergic" to the use of quenched and tempered steels for boiler and pressure vessel construction. World War II, however, brought about a change of attitude — and in quite a unique way. Studies of certain types of high-strength steels developed for armor plate for military use attracted the interest of members of one of the ASME subcommittees. These steels had many of the favorable characteristics sought for the construction of pressure vessels, most notably those designed for use in vehicles or other mobile units where weight was a factor. The Committee would eventually approve and write into the Code specifications and rules for using light-gage, high-strength quenched and tempered steels for pressure vessel construction, with a safety factor of three. The successful use of such materials during the war was reflected in the wording of Section III and Division 2 of Section VIII after the war was over.

Another provocative development during the war was the design and production of "packaged" boilers, units that were not exactly mobile but that could be easily moved to critical locations as needed. They met their biggest challenge, for example, in attempting to provide power in Europe in regions where the permanent power grids had been destroyed or severely damaged by enemy action. The four most common designs, all of which were of the bent-tube type, were cross-drum design, built for either stoker-or oil-firing; a portable SA type that resembled what had originally been designated for oil fields; and two "D" type units, one of which had been designed for ships that had supplementary steam requirements for heating and auxiliary power generation.

Packaged boilers were shop-erected steam generators that were shipped either fully assembled or in subassembly units that could be set up quickly and operated almost immediately with minimal erection requirements at the site. The demand for packaged boilers, which could be moved from place to place on short notice, had long existed, but was small and specialized. These units were most common in the oil fields, to meet extended or local requirements for steam and power. Before the war, portable fire-tube boilers first met that need and shop-assembled superheaters were perfected that were portable and separately fired to superheat the steam from these boilers for use with drilling rigs.

"The destruction of power stations and power networks in Europe and Asia during the recent war created a demand for standardized power plants whose components, including standardized boilers, would be available for shipment on short notice to any place in the world requiring power, and there assembled in accordance with the standard plans."

The demand was met by a series of emergency designs and units, variously referred to as "power trains," "unit power plants," "semi-portable power plants," "floating power stations," and "packaged power plants." After the war, the demand dropped off, yet the many designs and state-of-the-art applications contributed to the overall improvement of small boilers, and especially those for use in temporary power installations. They became valuable, for example, for use with industrial processing equipment that required considerable power inputs but were not readily accessible to permanent steam-distribution systems.

The earmarks of this kind of equipment, which naturally had to be reviewed for rules and specifications in the Code during this era, were compactness, portability, simplicity, and reliability. Compactness was essential so that each shop-assembled boiler could be transported as a unit (or in as few subassemblies as was practical) on standard railroad flatcars, truck beds, ships, or other carriers to sites that were often remote. These assemblies had to be sufficiently rugged to stand transportation on skids and the strain of being lifted into place by cranes and hoists. Simplicity was essential because in many instances the boilers had to be operated in regions — sometimes in foreign countries — where experienced manpower might not be available and where quick training of existing personnel was a vital factor. Reliability was obviously a major consideration, since the communities requiring packaged power were heavily reliant on continuous performance and since supplies of parts and the availability of experienced repair specialists might be all but nonexistent. In addition, these units often had to function at peak levels despite operational abuse, hostile environments, and low-grade fuels.

In a wartime atmosphere when an immense amount of technology was being focused on the design and production of more effective warships and faster fighters and bombers, the subject of steam locomotives had somewhat slipped from the public's mind. But the war had not been long in progress when they came into prominence as they took on the critical burden of hauling military supplies and serving as mass troop carriers. At the start of the 1940s, considerable attention had been paid to the performance of aging locomotives that had been taken out of mothballs or short-run schedules and placed on the main line. A mechanical engineer for the St. Louis San Francisco Railway Company, J. L. Ryan, pointed out a major flaw in the industry's treatment of its engines. "Locomotives are often built and maintained in kind for their service life," he protested, "renewal parts being made according to their *original* design, whereas at little — if any — additional cost, they might be renewed to modern design and proportions."

The demand for faster service, longer runs, and high mileage had left almost all of the railroads with many aging locomotives on their hands which were not adapted to meet such requirements. In simple point of fact, the horsepower demand could not be satisfied.

"From month to month," wrote Ryan, "one may read articles in railway publications that give the design features and proportions of locomotives being delivered to railways. If the reader is not something of a student of the motive-power field, he may come to the conclusion that the railways are being well stocked with new locomotives. This, however, is far from being true. In fact, time passes so rapidly, making obsolete locomotives that we are inclined to consider as modern, that those of us concerned with motive-power problems may well be startled by the actual conditions when making compilations of the locomotives handling our transportation services, their ages, proportions, and construction."

Assessing his own company as "an average-size railway," and applying relevant codes, standards, and power ratings as guidelines, the author estimated that only about 12 percent of the 610 locomotives owned "fully meet the transportation department's operating requirements and have the desired proportions for economy of operation and maintenance."

The complexity of the job facing the Boiler Code Committee can be dramatized by the problems that arise in a single industry with a single type of equipment: the American locomotive. The challenge became even more acute during the Second World War when consideration had to be given to serious shortages of materials and equipment, as well as to fuels.

"When the credit for winning World War II is apportioned a large share will be given to the coal-burning steam locomotive," wrote C. F. Hardy, Chief Engineer for Appalachian Coals, shortly after the end of the war. As he pointed out, the task of the railroads was all the more gargantuan because of the disparity in wartime supplies of coal, whose available types and sizes were often vastly different from conventional supplies prior to Pearl Harbor. Lumps of coal, which most people picture as coming in half a dozen — certainly no more than a dozen — different sizes, were actually being supplied in 252 different size designations, according to one industry report.

While codes and standards do not cover basic fuels, it was pointed out, oftentimes the availability and quality of fuels plays a vital role in the nature and degree of certain rules and specifications for equipment using those fuels. In the case of locomotives, this has always been true — in war and peace alike

— complicated by the fact that boiler and engine parts are subjected to constant stress because of the pressures, temperatures, rapid motions, and joltings to which they are constantly subjected.

The Second World War prompted a searching review of the nature, purpose, and use of the ASME Code Symbol Stamp. It had long since been established quite clearly which committees and organizations had jurisdiction over the use of the Code Stamp and its various imprints. As was mentioned earlier, the very first Code, devised in 1914, had provided for a Code Symbol Stamp to identify equipment designed, fabricated, and inspected in accordance with Code rules. The symbol had been created so that accredited manufacturers could certify that their products had been designed and made in accordance with ASME Code rules. However, with the passage of time, the roles of those involved became somewhat cloudy. This was particularly true during World War II for two specific reasons. First, many applications of civilian boilers, pressure vessels, and related components no longer came under the ASME Code once they were assigned to military installations and uses. The Army and Navy did, of course, have their own rules and specifications, many of which had been greatly influenced by the ASME Code, but these were much more flexible under emergency conditions and certainly battlefront operations when equipment had to be used no matter how many flaws or weaknesses might be evident or suspected. Second, even when the Code had jurisdiction over equipment, there was greater flexibility in many instances when the users came under government or military control or when conventional inspections and reports were difficult to continue on a regular basis. So the question "Whose Stamp is it anyway?" was heard, in one way or another, over and over again.

Unfortunately, the end of the war did not resolve the question, at least not entirely. As recently as the mid-1980s, the question was still echoing, despite the fact that the relationships between ASME, manufacturers, inspectors, and users had long since been well defined.

Since the inception of the Code, protection of the integrity of the Code Symbol Stamp had been of primary importance. A critical point had been reached in 1941 when the United States Justice Department inquired into the procedures of the Boiler and Pressure Vessel Code Committee. This investigation resulted from complaints charging several insurance companies with refusing to offer inspection service to some manufacturers, an attitude that prevented the latter from producing Code products and components.

The investigation revealed that the insurance companies so cited were actually withholding their services from certain companies which did not have the means to produce vessels that could meet Code specifications. As it turned out, these insurers were simply assisting ASME in protecting the integrity of the Code Symbol Stamps. In order to discuss the results and the implications of this investigation, the Code Committee called two special meetings. The discussions affirmed and reaffirmed the idea that it was of paramount importance for the Committee to retain the responsibility for issuing and protecting the Symbol, and not simply to restrict its activity to technical matters. To that end, strengthening revisions were made to the provisions for stamping and inspection, clarifying enough to induce the Justice Department to drop its investigation and accept the stipulation that the Committee should retain the responsibility for protecting the integrity of the Code Symbol Stamp.

This was a major milestone in the history of the Code because it established the objectivity of the Code Committee's whole system of accreditation and supervision, through which it could provide a clear-cut means for distinguishing conforming Code equipment.

Over the years the Code Stamps have evolved from just a single "modified four-leaf clover" imprint to 18, each one covering a different subject area under the 11 Sections of the Code. Most were established to cover specific products that had finally assumed importance in the Code structure. Some, however, came into being because of special situations that developed and that did not seem to be adequately covered by existing Stamp Symbols. A case in point was the "PP" Stamp, which first made its appearance in Section I of the Code in the 1940 Edition. This Stamp was created by the Boiler and Pressure Vessel Committee in response to an application by a Philadelphia manufacturer that intended to erect boilers without actually manufacturing any parts of the boilers.

The Committee had first considered the firm's application in 1937 for an "S" Stamp. The application was rejected since "it was the Committee's opinion that, in order to assume the responsibility for boiler construction implicit in the 'S' Symbol of the period, it was necessary for the 'S' Stamp holder to have manufactured a major part of the boiler."

Persistence on the part of the company to obtain a stamp to which it felt entitled, led to the appointment of a Special Committee on the Issuance of ASME Code Stamps. The lengthy reviews and deliberations that followed eventually led to the creation of the "PP" Stamp just before World War II, making it possible for a company to obtain this kind of accreditation and be responsible *only for work performed,* while the manufacturer(s) of the components assumed the ultimate responsibility for the parts produced and turned over to the erector.

The move was made at an auspicious time, for with the advent of the war, numerous situations began to arise in which the company erecting a boiler or other unit was not the producer of the unit or all of the components. This action in regard to the "PP" Stamp represented a significant guidepost in the history of the Code because the Committee had now recognized and established the concept that "design responsibility must be taken for all aspects of Code construction."

A further review, right after World War II, resulted in the issuance of the "A" Stamp, mainly of use to holders of the earlier "S" Stamp or the "PP" Stamp described above. The Committee's main purpose was to clarify the Code further regarding the responsibilities of manufacturers or-contractors for qualifying the welding completed by their organizations.

Walter Harding, after many years of working on Code committees and dealing with related problems both internally and externally, interpreted the policies regarding Code Symbol Stamps in a statement made expressly for this history. "The Boiler and Pressure Vessel Code is a set of rules for the design and construction of vessels which will be safe in resisting the expected service loads, primarily pressure," he wrote. "The Code Symbol Stamp is applied by the manufacturer as his certification that the product has been built in accordance with applicable Code rules. The Code Symbol Stamp thus applies to the *new construction* of such products. Delivery to, and installation in, the purchaser's (user's) facility basically ends the product manufacturer's responsibility and the applicability of Code rules. Operation and maintenance of the Code-stamped equipment become the responsibility of the user and may be subject to laws of the jurisdiction in which the equipment is installed and operated. Conditions which develop in service must be evaluated on an individual basis among the user, his insurance carrier, and the enforcement authority of that jurisdiction.

"The rules of the ASME Boiler and Pressure Vessel Code are intended to provide for the design and construction of equipment which will be safe under the specified pressure and temperature conditions. The Code does *not,* however, specify how the equipment is to be designed to achieve its intended performance as part of an operating system. The Code does not specify the size of a drum or pressure vessel or the size and number of nozzle openings. After the designer has selected these items (and materials) to suit the functional needs of the system, the Code rules tell him the minimum thickness of the pressure parts. Similarly, the number, size, and arrangement of tubes to accomplish a desired heat-transfer function must be established by the designer without guidance from Code rules. The Code then tells him the minimum thickness of such tubes depending upon the selected tube size and material, the specified design pressure, and the designer-calculated tube metal temperature.

"This important distinction means that manufacturers *compete* in performance-related aspects of equipment but do not compete in such pressure—safety related items as shell thickness formulas or allowable stresses. This explains why engineers representing different manufacturers can serve on the safety-related, common-interest activities of the Code without concern about anti-competitive or antitrust liability. The competitive aspects of equipment performance are not discussed in the Code Committee meetings, nor are they made the subject of Code rules."

With all of these considerations taken into account and in light of the Code's history of accomplishment, it was not surprising that special recognition would be given to the Code by the American Standards Association (later to become ANSI, the American National Standards Institute). In 1952, ASA conferred the following citation on the ASME Boiler and Pressure Vessel Committee:

"Probably no other standard in America has done more for national safety than the ASME Code."

CHAPTER 15

THE NUCLEAR DAWN

A New Era
and the Aftermath of World War II

In a book entitled *Wonders of the World,* published in 1959, the British nuclear power plant at Calder Hall was listed as "one of the seven wonders of the modern world." Although not the first nuclear energy reactor, it was the world's first large-scale electric-generating station powered by atomic energy to supply power for industrial and domestic purposes.

Located in West Cumberland, England, it was established by the United Kingdom Atomic Energy Authority and officially went on stream in 1956, designed for the dual purpose of generating electricity and producing plutonium for defense. The Calder design was the prototype of other such nuclear power stations constructed in the United Kingdom and America to replace those that were conventionally fueled by coal and petroleum. The facility was a peacetime symbol of what would take place when the tremendous energies that had been directed to the production of materials of war were freed for a new purpose. In order to fuel the burgeoning new economy that had emerged and provide the comforts and products expected by the public, the engineering profession was faced with brand new challenges — ones that not even the wartime crises had thrust upon them. The epicenter of this enormous growth was *power*. In the field of boilers alone, new designs were needed as turbines of larger capacity came into being. The extent of the change was evident in many ways, most dramatically perhaps in the fact that postwar turbine-and-boiler combinations could singlehandedly produce several times as much electricity as was possible with an entire pre-war generating station.

An important milestone in the progress toward the production of electricity at the lowest possible cost was the successful operation in 1957 of the first commercial operation of the "supercritical" steam generator at the Philo Plant of the Ohio Power Company on the American Gas and Electric System. This functioned on the theory that water in a boiler, under the "critical" pressure of 3,206 pounds does not bubble or boil as it converts to steam but changes instantaneously from water to steam at a temperature of 705°F. This type of forced-flow, once-through boiler, which needed no drum for collecting steam and water and was composed of independent, continuous tubes, could be operated both above and below critical pressure.

"We all conceded that nuclear power and the process of its generation implied a sufficient number of side dangers to merit broad consideration," recalled Frank Williams, who had participated directly in many of the ASME activities leading to nuclear codes and standards. "We realized fully at that stage that nuclear equipment had to be designed with meticulous care, that innumerable details had to be considered, that inspections would require great care, and that the competence of the people in every phase of the operations needed to be unfailing in order to understand the differences and the consequences of any and all kinds of malfunctions.

"Today, the facts of life are that we have a great number of nuclear plants operating effectively that we never hear anything about. But ASME took the position through its Code Committee that we were going to do these things. And so there are a whole series of requirements, starting from the beginning and going through the qualifications of inspectors and testing that are put there to provide the extra degree of safety felt to be necessary to protect against the extra degree of risk. From that point forward, the basic engineering is nothing unusual.

"In the early days of nuclear involvement, we at Taylor Forge, along with a number of other companies, such as Kellogg and Westinghouse, made a lot of equipment in the form of pipe fittings, piping, nozzles, and the like for an experimental plant and then for the very first nuclear power plant at Shippingport. So we had an insight as to what probably would happen. At first, we talked informally about this subject, and then proposed the course of action we felt we should all take. The result was the formation of a study committee, whose initial assignment was to scan what was going on and make a report. The members were basically involved in the *system,* rather than the mechanics of nuclear energy.

"These activities resolved themselves at a meeting in Tulsa, Oklahoma, where a decision was reached that a separate subcommittee should be established to look at the matter and make recommendations. We tried to balance the committee with members familiar with the systems and other members who knew the mechanical side. The chairmanship of that committee was assigned to me, as someone intimately aware of Code policies, which position I held until the work was fairly well under way. It was quickly recognized and agreed upon by our members that whatever rules and specifications ultimately evolved they should be compiled in a separate section of the Code. Since Section III — formerly Locomotive Boilers — had become defunct, that was then assigned to the nuclear document.

"Because of many reasons that are hard to define but which we felt to be valid at the time, Section III provided a greater degree of leeway than was characteristic of the other sections and, in effect, became almost an independent Code. Looking back, we could see that much of what occurred was an outgrowth of earlier Code Basis work.

"We all worked very closely in those years with the Atomic Energy Commission and the Nuclear Regulatory Commission. Close coordination was established, with the government agencies being willing to accept our rules, yet reviewing them and always reserving the right to require something more stringent. One of the really important facts about nuclear was that it led us, within ASME, to develop a positive set of requirements for inspectors and a sound system of qualification for inspectors. Also, it motivated the establishment of a certification program which did not delegate to the insurance companies or the states or the federal government, the qualifications for the nuclear stamp. This turn of events took us into a whole program that had never functioned quite like it before.

"In everything that was planned or accomplished, quality assurance was the key phrase."

Power engineers, taking some lessons from World War II when a great deal of research and development had been aimed at learning how to conserve critical materials, were now directing their efforts also at ways to utilize the energy in waste gases and byproduct fuels. It had long since been demonstrated that boilers could be specially designed to convert available energy into useful steam, most notably in energy-related industries. In the petroleum industry, for example, catalysts used in cracking plants would quickly become coated with carbons that had to be burned off in order to maintain the efficiency of the processing operation. Since the removal of the carbon resulted in considerable quantities of waste in the form of carbon monoxide (CO), the question arose: Why not use this waste gas as a boiler fuel? Through a refinement of this effort to utilize waste, it was discovered that the saving in fuel alone would, within two years, almost pay for the purchase of sophisticated new boilers that could use the byproducts to produce steam.

Later, this kind of waste utilization also paid off by avoiding many of the problems of pollutants rising into the atmosphere, thus preserving the environment and avoiding the huge expenses associated with fighting contamination.

Nuclear power was not a phenomenon that arrived overnight at the end of World War II. Its practical origins were relatively short, considering the immensity of this new source of power. By the middle of the 1950s, private industry had become heavily involved in nuclear development. The first privately financed plant for the manufacture of atomic energy elements was placed in operation early in 1956, later supplying such products as fuel elements for research and test reactors and cores for power reactors.

The steadily rising increases in temperatures and pressures for boilers and other power equipment in the 1950s demanded, among other things, heavy-wall forgings suitable for advanced designs in pressure vessels and piping. The hydraulic press came into its own in this field, for the formation of the heavy plate required for high-pressure boiler drums and nuclear vessels. Besides the demand for extremely heavy-walled and large-size piping, the industry had to turn its attention to the specialized needs not only of atomic energy but the advances in jet propulsion, petroleum refining, and chemical processing. These

165

requirements led to more sophisticated specifications for such products as tubes for boilers, waste heat recuperators, and heat exchangers. New alloys also came into the picture, such as ferritic stainless steels and blends of titanium, zirconium, and molybdenum.

Advancing rapidly, too, were refinements and improvements in producing seamless piping, piercing and drawing tubes, and using extrusion to fabricate metals that could not be produced by other means because of their metallurgical characteristics. Cold-finished tubing was also made possible by innovative machinery that could reduce hot-formed tubes to size by using compression rather than heat or drawing. In addition, a number of ingenious ways were perfected for producing solid shapes through methods other than conventional rolling, and the introduction of high-speed welding. Seamless welded fittings were extensively in demand for piping used for steam and in the refinery and processing industries.

The significance of such developments to the ASME Code was that these new methods resulted in materials and products that were less susceptible to flaws and weaknesses and that, most importantly, could be relied upon to meet much closer tolerances. In trying to understand and cope with unfamiliar developments in the field of nuclear power, members of nuclear-related subcommittees investigated ongoing projects that were pioneering in nature. One example was the Vallecitos boiling water reactor, the first privately owned and operated nuclear power plant to deliver significant quantities of electricity to a public utility grid. Located near Pleasanton, California, west of San Francisco, this plant delivered approximately 40,000 megawatt-hours of electricity during the period October 1957 to December 1963- (For this achievement it was designated an ASME Landmark in 1987.)

This reactor, a light-water moderated and cooled, enriched uranium reactor, used stainless steel-clad, plate-type fuel. It served as a pilot plant and test bed for fuel, core components, controls, and personnel training for the Dresden Project, a Commonwealth Edison plant built in Illinois five years later. The plant was a collaborative effort of General Electric and Pacific Gas and Electric, with Bechtel serving as the engineering contractor.

In 1955, the chairman of the Boiler and Pressure Vessel Committee appointed a Special Committee to Review Code Stress Basis. "The charge to this committee," explained Leonard P. Zick, "was a complete review of the existing Code requirements balancing material properties for all materials, design approach and details, construction, examination, pressure testing, and inspection. One of the main discussion points was the European pressure to design on yield only. Material properties included those with low yield-to-ultimate ratios and those with high yield-to-ultimate ratios. Toughness, heat treatment and effects of welding were others."

166

During several years of discussion, recalled Walter Harding, "this Committee began to evolve the concept that higher allowable stresses could be permitted, with no reduction in safety, provided that (1) a much more detailed design analysis of stresses could be achieved, including the effects of combined stresses, secondary stresses, and peak stresses; (2) more stringent requirements could be placed on materials, including nondestructive testing to evaluate resistance to brittle fracture and other conditions; and more stringent requirements could be demanded on fabrication processes, particularly welding, heat treatment, and nondestructive examination of the vessel."

He related that, during this same period, a separate subcommittee on nuclear power had been studying the proper role for ASME in this rapidly developing field. This subcommittee came under increasing pressure from the Atomic Energy Commission (AEC), later renamed the Nuclear Regulatory Commission (NRC) to formulate a safety code for construction of nuclear pressure vessels. "Such a code," he said, "had great need for higher allowable stresses to keep the wall thickness of the large-diameter, high-pressure vessels to a practical value. At the same time, the anticipated operating conditions of nuclear components — which could impose thermal, fatigue, and shock loadings on the basic pressure load — called for a much more detailed and extensive design analysis."

By the end of the 1950s, the Main Committee recognized that the concepts evolved by the Special Committee to Review Stress Basis matched the need for nuclear components. The Special Committee was then instructed to expedite its effort toward devising an ASME code for nuclear components, an action that resulted in the first edition of a Section III.

The Special Committee then reverted to its original charge and compiled an advanced code for non-nuclear components that was later adopted as Division 2 of Section VIII This became the added responsibility of the Subcommittee on Unfired Pressure Vessels and the Special Committee was discharged.

Nuclear construction required more quality control than did other fields, though it was largely an internal program, to satisfy manufacturers themselves. Quality assurance required reviews by outside parties and the question arose as to the best ways of qualifying inspectors (third parties) for nondestructive examination, for example.

Speaking on the subject of nuclear certification in January 1977, Melvin R. Green, Managing Director of the ASME Codes and Standards Department, explained that "in 1965, ASME included nuclear vessels in its Certification Program. A certificate was issued based on a favorable report from the authorized inspection agency and the jurisdictional authority. Industry and government wanted more! On July 1, 1968, the concept of nuclear survey teams became mandatory. Since that time, Section III of the Boiler and

Pressure Vessel Code has been expanded to cover piping, pumps, valves, concrete reactors and containments, core supports and component supports. The organizations producing these items that are presently addressed in the Code are manufacturers, constructors, fabricators, engineering organizations, installers, and owners. An ASME Survey Team conducts the nuclear surveys under the direction of an ASME Team Leader with the participation of an ASME consultant, National Board representative, representatives of an applicant's authorized inspection agency, a utility representative, and a representative of the jurisdictional authority."

In 1975, the PBCS established a NPCS Supervisory Committee to administer ASME's Nuclear Power Codes and Standards activities, with the responsibility to assess the needs for nuclear power codes and standards and related certification programs; establish the necessary committees to develop standards; define the parameters for certification programs; monitor committee activities; consider proposed nuclear standards for ASME approval; provide for due process; and keep the PBCS advised of the status of all activities related to nuclear power.

As Green further explained, "The purpose of the ASME nuclear certification program is to ascertain that the rules of Section III of the Code dealing with Quality Assurance are a part of the applicant's control system so that the applicant has, and uses, a document known as a quality assurance manual. The document must be clear and understandable so that it can serve as a working guide."

As a result of requests from industry and government, ASME later broadened its nuclear accreditation program to include material manufacturers and suppliers. Among other actions, it also developed criteria for the qualification of consultants, developed standards for authorized inspectors and inspection agencies, and provided for the periodical publication of lists of nuclear certificate holders.

As in other spheres of activity, Code Symbol Stamps stood as the hallmark of acceptance and certification. In August 1966, a Task Force convened in Erie, Pennsylvania, to consolidate views on the principles involved in the inspection and quality control of nuclear vessels. Taking into account the fact that it could not be assumed that nuclear inspectors would all be specialists in this field or that they would even be experts in every field of nondestructive testing, the Task Force reached several conclusions. Among these were the necessity of providing inspectors with pertinent and detailed facts about equipment to be tested; the establishment of standards for nondestructive testing; the institution of training programs at practically all levels of industry to familiarize participants with processes and equipment; and tightening up the acceptance criteria on manufacturers before issuing nuclear stamps.

At this time, there were 67 holders of "N" Symbol Stamps in the United States. In 1967, the Committee devised a questionnaire to be completed by all future applicants for the stamp and the Certificate of Authorization for its use. The questions covered such subjects as scope of fabrication intended, availability of a professional engineer qualified for the certification of stress analysis reports, background of the manufacturer's staff with respect to nuclear pressure vessels, familiarity of the applicant with ASME Codes (particularly Sections III, VIII and IX), nature of an existing quality control program, availability of instruments for hydrostatic and other tests, and, of course, details about personnel, facilities, and equipment,

The situation relating to Code Symbol Stamps had already been complex, long before nuclear technology had to be reviewed for Code application. So it was evident from the start that the Committee was faced with an extraordinary challenge. The insurance companies had perfected systems of inspection and trained inspectors, basically in their own defense to minimize the number and extent of claims. State governments also became involved because they had to pass laws governing inspection procedures. The Boiler and Pressure Vessel Code Committee was obviously concerned because it had to determine whether the Code rules were actually being adhered to — a situation that had led directly to the creation of the Code Symbol Stamp. Who would be qualified and permitted to apply the stamp?

The debates became endless and often controversial and heated. The states took the position that they had the final say since they were enacting the laws. The insurance companies felt that this was a duplication. And ASME was caught in the middle between the state regulators and the private inspectors (who in turn were governed by state regulations). What it ultimately boiled down to was that the symbol and stamp would be controlled by ASME, which was responsible for the compliance of those who received authorization and whose incentive was certainly to retain their usage rights; the state, through its inspectors who belonged to the National Board, would accept the certification of inspection; and the insurance companies would qualify their own inspectors through the completion of the proper training and tests.

Prior to 1968, ASME had depended on the jurisdictional bodies or inspection agencies for recommendations relative to the qualifications of applicants for accreditation to use Code Symbol Stamps. Then, in July 1968, more comprehensive Code requirements were put into affect regarding applicants for nuclear accreditation. These requirements introduced Quality Assurance on a more formal basis and also initiated the aforementioned nuclear survey teams. Since then, requirements in other sections of the Code evolved to acquire a review team; the revisions have maintained the principle that an authorized inspection agency must have a potential legal or insurance interest in the finished product to be stamped with the ASME Symbol. The inspector must assure himself that the manufacturer conformed to the Code rules. The National Board acknowledged the value of this organizational procedure for reviewing Code Stamp applicants and began to participate in the survey teams in July 1968.

These new requirements were all built upon a history of success and achievement, accepting the solid foundations of the past yet, at the same time, recognizing the complexities of advancing technology and changing public interests. Leonard P. Zick, explaining the coverage of the Code and participation by third parties who bring objectivity to the procedure, observed that there were two main areas to consider. One was the non-nuclear construction of boilers and pressure vessels. The other was nuclear power, which he pointed out covered a great deal more hardware. In addition to the reactor vessel itself, the Code had to consider the containment vessel, heat exchangers, pressure piping, tanks, and many other components.

"A fabricator can build an open-top oil storage tank for the process industry," he explained, "and construction will be governed by the American Petroleum Institute standards. Build the same tank for water storage and install it at a nuclear site and its design and construction will be covered by Section III of the ASME Code. During the construction phase of a nuclear power plant, inspections are also performed by the United States Nuclear Regulatory Commission (USNRC) as part of the licensing requirement.

"The special committee on nuclear power was organized by ASME in 1955. Section III was first published eight years later. The 1971 edition of Section III marked the transition of this code section from a vessel code to one covering the entire system involved in the conversion of nuclear power to steam power. It was followed by the adoption of Section VIII, Division 2. Both sections had safety factors of three.

"Interestingly, there is a difference in terminology in some instances between descriptions and specifications for non-nuclear work One example is the use of the phrase 'quality control' for non-nuclear applications. When applied to the nuclear field, it becomes 'quality assurance.'"

Maintenance in a nuclear power plant involved methods and equipment necessitated by the requirements of radiological safety, quality control, inspection, and operational testing. It was found that the most frequent maintenance problems affecting the safe and efficient operation of a nuclear plant related to primary coolant leaks, instrumentation and control equipment failures, and steam equipment malfunctions. Other maintenance problems unique to nuclear plants were found to be auxiliary-system-fluid leaks that required reactors to be shut down, the inaccessibility of key components while reactors were in operation, reactor containment integrity to assure the isolation of leaks, and the very size and complexity of reactor coolant pumps.

Steam generators for nuclear plants were actually tubular heat exchangers, functionally quite similar to their fossil-fuel counterparts. However, since the inside of the tubes contained the reactor coolant, all of the planning for inspection and maintenance had to take into account the possibilities of radioactivity. As experience was accumulated, it was seen that the radiation levels in the tube sheet areas were enough so that inspectors and workers could not remain in the vicinity for more than an hour without being exposed to dangerous levels of radiation.

Orientation programs assumed an importance never before experienced by ASME subcommittees in the nuclear field, even though Code personnel had a distinguished record of diligence and application in the matter of homework. Among thousands of other details, they learned that nearly all of the power reactors in the United States were "water reactors," using ordinary water as the reactor coolant. They fell into two categories.

— The *pressurized* water reactor (PWR) in which the cooling process is done by water under considerable pressure, wherein the high-pressure water is conducted to heat exchangers and steam is then produced on the low-pressure side of the heat exchangers, which are commonly referred to as "steam generators."

— The *boiling water* reactor (BWR) in which the coolant water is allowed to boil in the reactor, and in which the steam is generally transmitted directly to a turbine for electric power generation.

From the inception of nuclear power plants, it was widely recognized that *safety* had to be a prime consideration not only because of potential radiation hazards but because nuclear chain reactions were known to be capable of enormous power transients if not inhibited by design and properly contained. Thus, the priorities fell into line: stable and reliable materials, technical and physical control, heat removal, and structural integrity.

In order of priority, materials certainly were of prime consideration, since only a few offered the necessary strength, stability, and other properties that would be compatible with both the physical demands and the unique environment. Four materials were, after endless rounds of research and testing, found to be essential in the design of a pressurized water reactor that could be utilized economically in the production of power.

- *Pure coolant water with good heat-transfer characteristics,* which possessed the additional attribute of reducing neutron energy and thus enhancing neutron absorption in the fissionable material.

- *Uranium dioxide,* a highly stable material that was insoluble in water, withstood irradiation well, and could be loaded easily into fuel rods.
- *Zirconium alloy,* which had reliable strength and low neutron absorption and was sufficiently resistant to interaction with hot water so that it could be used for cladding tubes for reactor fuel rods.
- *Steel,* selected as the metal for use in the pressure vessel that contained the reactor. Stainless steel, too, was found to be ideal for lining the pressure vessel to minimize corrosion, for thermal shielding, for reactor supports, and for core structure, where zirconium alloys lacked the required strength.

All of these materials and factors, and many more, were eventually covered by the Code as nuclear power became a reality and research defined the parameters and established the specifications for materials and components in exhaustive and complex detail. Auxiliary systems also became a vital part of the overall plant makeup. In addition to the main units of the nuclear steam supply structure for instance, typical auxiliary systems were designed that would reliably process purification and reduce impurities, chemical addition and sampling, waste disposal, intermediate cooling of numerous components, spent-fuel cooling, and decay-heat removal following the shutdown of a reactor.

The Congress of the United States has vested in the Nuclear Regulatory Commission (NRC) responsibility for the safety regulations of nuclear facilities with respect to radiation hazards. This control is accomplished by NRC rules, standards, instructions, and licensing mechanisms to govern the use of nuclear material and nuclear facilities to minimize danger to health and property. All reactors must be licensed by the NRC and in the case of power reactors, it is necessary to obtain a construction permit and, prior to operation, an operating license, customarily granted on a provisional basis. After a period of successful operation, a long-term operating license (up to 40 years) is issued. Yet the reactor is always under the regulatory surveillance of the NRC.

In order to obtain a permit for the construction of a nuclear facility, the applicant (who must be the prospective operator) must file with NRC a formal application including a safety analysis of the proposed plant. This is reviewed formally by the NRC's regulatory staff. The Advisory Committee on Reactor Safeguards (ACRS) conducts an independent review and makes its report to the NCR staff. The ACRS report and the comprehensive NRC staff hazards evaluation are published and distributed to appropriate state, local, and federal officials and to the press and general public. A public hearing is scheduled and announced at least 30 days in advance by *Federal Register* and other publications, and letters to officials. The public hearing serves as a forum in which any issues relating to the nuclear facility can be aired and evaluated. After the hearing, the Atomic Safety and Licensing Board reaches its decision. If there is no appeal, the NRC commissioners review the decision and either approve or reverse it.

Given final approval, the operator of a nuclear plant must obtain an operating license, which specifies among other things the maximum power at which the plant may be operated. A set of technical specifications defines other operating limitations that must be observed.

The internal components of reactors are designed to withstand the stresses resulting from start-up, operation, and shutdown conditions. Reactor internals are fabricated from materials designed and approved within the allowable stress levels permitted by the ASME Code, Section III, for normal reactor operations and transients. Structural integrity of all core-support-assembly circumferential welds is assured through compliance with Section III or Section XI, or both, radiographic inspection acceptance standards, and welding qualification. All pressure-containing components are designed to meet the rules and specifications of Section III.

Lest it be mistakenly thought that nuclear power dominated the postwar period, attention is drawn to other scientific and technical areas that were vital to the Code and in which major achievements were recorded. One such area covered pressure vessels and piping. New criteria of substance were formulated by the Code Committee in the 1950s. "It was during this period," wrote R. L. Cloud in the introduction to an anthology on PVs and piping, "that a reasonable understanding of brittle fracture and low-cycle fatigue was reached, the principles of limit design were established, and the mechanisms of thermal ratcheting, shakedown-to-elastic action, and constrained plasticity were explained."

Of equal importance to the understanding of physical processes were the advances made in computational methods. The computer performed the dual accomplishment of making the solution of complex problems a matter of routine and at the same time keeping the procedure economical.

By the 1950s, the Code had achieved "a dominant influence on pressure vessel design in the United States." Thus, when the design criteria for pressure vessels were set forth in Section I, Power Boilers, and Section VIII, Unfired Pressure Vessels, they were readily accepted and acknowledged as the culmination of the accumulated experience of many years of pressure vessel design.

"They were successful in establishing a favorable safety record in industrial practice." Nevertheless, stated Cloud, "the success of the Code resulted not so much from a consistent philosophy of design but from the fact that experience and sound judgement formed a part of all the individual features of the criteria."

The closest relationship to any general rules that could be distinguished in the Code was that, first of all, it addressed itself to maintaining overall stress levels at low values and, secondly, that it required the use of ductile materials that could safely accommodate local peak stresses and discontinuity stresses. Cloud pointed out that a measure of the progress in the development of design criteria during the 1960s was that by then there were four criteria for pressure vessels and that they differed markedly from the two criteria that had existed in the 1950s.

"The new criteria are based on fundamental considerations of analysis and material behavior," he explained. "There is a unifying philosophy of design in the sense that an attempt has been made to understand all possible modes of failure and provide rational margins of safety against each type of failure in a manner consistent with the consequences of that type."

Several basic concepts in regard to pressure vessels emerged as a result of this renewed quest for fundamental understanding.

- Since thermal stresses were explicitly considered and applied, vessels built to the new criteria had a balanced design that carried thermal and mechanical stresses much better.

- The basis for design stress limits and structural evaluation was shifted from the maximum principal stress theory to the more accurate maximum shear stress theory, a fundamental change that was very important in regions where one principal stress is tensile and the other compressive.

- It was recognized more fully that pressure vessels were subject to fatigue failure, especially those that were exposed to changing thermal stresses, making it possible to define specific procedures for detecting and evaluating fatigue damage.

- Brittle fracture as a mode of failure controlled by temperature limits and design details, rather than stress limits, had become much better understood, leading to methods for detecting, evaluating, and preventing this specialized problem.

- Studies of failures in ductile material after some plastic action resulted in new concepts of elastic analysis and an application of plastic theory that was a distinct departure from past practice. It was seen as a more clear-cut understanding of fundamental mechanical behavior, made possible by major advances in applied mechanics theory and computational ability.

"The dominant feature of the revolution in design criteria," concluded the evaluation, "is the concept of *design by analysis*," which by this time in the history of the Code had become the method required for all Section III vessels, as well as the basis for the philosophy motivating the standards of Section VIII, Division 2. The late 1940s and early 1950s were segmented as a period in which considerable strides were made in understanding and combating corrosion and related forms of deterioration. The cause of the so-called "brittle boiler tube" or "plug-type corrosion" which had occurred in the higher-pressure steam generators (around 1,300 psi) had been explained by many theories, which had not all been in agreement in accounting for this type of metal attack. One theory involved the reaction of copper with steel; another assumed that the so-called embrittlement resulted from the action of concentrated sodium hydroxide formed in a film boiling at the tube surface; while a third explained the action on the claim that dissolved oxygen in the feed water was the offending agent.

F. G. Straub, conducting engineering research at the University of Illinois, who wrote this paper on wall-tube corrosion in steam-generating equipment operating around 1,300 psi, concluded that "dissolved oxygen in the absence of an oxygen scavenger" was the main cause of this type of failure. Neither copper nor film boiling were the culprits.

Another type of failure was described as "barnacle corrosion" and metal embrittlement. A characteristic example was that of three boilers (of 1,350 pounds operating pressure) that displayed these flaws in 1945 after seven years of operation. The "barnacles" formed on the tubes and multiplied until one actually blew out, leaving a hole and causing the equipment to fail.

In a 1952 report, "Analysis of Some Corrosion Problems in Petroleum Refineries," J. F. Mason, of the Corrosion Engineering Section of International Nickel, described his experience over 25 years. "Corrosion of equipment in petroleum refineries for many years has represented a major portion of operating costs, and oftentimes this condition has been aggravated by the misapplication of available metals and alloys."

In order to test alloys, researchers devised a "spool holder" into which could be inserted a dozen or more discs, each one composed of a different alloy. Previously cleaned and weighted specimens were mounted on the spool with nonmetallic brackets of porcelain or bakelite to separate and insulate them from each other, as well as from the metallic parts of the holder. Two similar specimens of each material were included in each spool. The completed test assemblies were then fastened firmly in place in the desired test locations inside the operating plant equipment, where they were allowed to remain for sufficient periods of time to provide reliable indications of corrosion behavior.

175

Each specimen was carefully weighed and each had exactly the same area of exposure, 0.5 square decimeter.

Upon completion of the tests, the specimens were photographed and examined to determine the nature and extent of the corrosion, cracking, pitting, or other forms of local attack in each case. They were then cleaned of all scale and corrosion and weighed again. From the weight losses, areas of the specimens, and duration of exposure, the corrosion rates could be precisely calculated in terms of milligrams per square decimeter per day. These rates were then easily transposed into inch penetrations per year. Where pitting or local attack had occurred, the depth of the deepest pits was measured microscopically or with a depth gage.

It was concluded from these tests that, in numerous cases, the most appropriate alloys were not being used and the solution was simply to make replacements. One example was that of steel tubes in a heat exchanger in a petroleum refinery. After pitting and general corrosion began to occur, the tubes could be expected to last only nine months before failure would occur. By replacing the steel with Muntz-metal (an alloy containing copper and zinc), the life of the tubes could be extended by 50 percent or more.

Another subject of steadily growing significance in the development of the Code related to tube formulas. Under the direction of Walter Harding in the early 1960s, painstaking studies were conducted to compile data about tubular rupture data on thick-walled tubes. Tubular stress-rupture tests proved that when the stress in tubular specimens was evaluated under the 1962 Code formula, the actual life of the tubular specimens did not always agree with the rupture life of the bar specimen at the same stress. The discrepancies became even greater as the ratio of thickness to diameter increased. These findings led the task group to obtain additional data on other materials and to give all of this information a more detailed evaluation. There had been no instances known to the task group in which tube failures in operating boilers had directly been attributed to the formula itself, without other factors. These evaluations resulted in the group's recommendation that a change in the formula was justified, at least enough so the expected rupture life of a tube without thermal stress would equal that of a uniaxial bar specimen.

It was common practice to continue operating boilers when there was a small leak in one of the tubes. This procedure was warned against because it had been found by experience and testing that steam or water escaping from even a small leak could cut other tubes by impingement and set up a chain reaction of tube failures. Also, because of the loss of water or steam, a tube failure could alter boiler flow or circulation and result in the overheating of other circuits and components. "This is one reason," it was admonished, "why furnace risers on the once-through type boilers should be continuously monitored. A tube failure can also cause loss of ignition and, if re-ignition occurs, a furnace explosion."

Any unusual increase in furnace-riser temperature on the once-through type of boiler could be an indication of furnace-tube leakage. Small leaks could sometimes be detected, it was found, by the loss of water from the system, the loss of chemicals from a drum-type boiler, or noise made by the leak itself. The only alternative in these cases was to shut down the boiler as soon as normal operating procedures permitted and locate the trouble through hydrostatic testing.

Finally, there was the matter of the development of steels for low temperature service in the late 1940s and early 1950s to meet the demand for the increasing use of pressure vessels at low and cryogenic temperatures. Prior to the Second World War, the Code had not considered environments for pressure vessels below -15°F, so the situation called for the development of a whole range of new steel alloys and specifications. It was fortunate that the austenitic stainless steels, originally perfected in Germany prior to World War I, were already included in the Code and were strong and tough down to hostile temperatures of -300°F or -400°F. These steels proved to be very useful for use at cryogenic temperatures, but were quite expensive. New steels, mostly employing nickel for an alloying element, were developed for use from -15°F. to as low as -300°F.

With all of these innovative programs in existence, not to mention the complexities of the codes and standards for nuclear power plants, it was evident that this era was a trying and challenging one for the Code Committee.

CHAPTER 16

HISTORY REPEATED

Full Cycle into the Unfamiliar

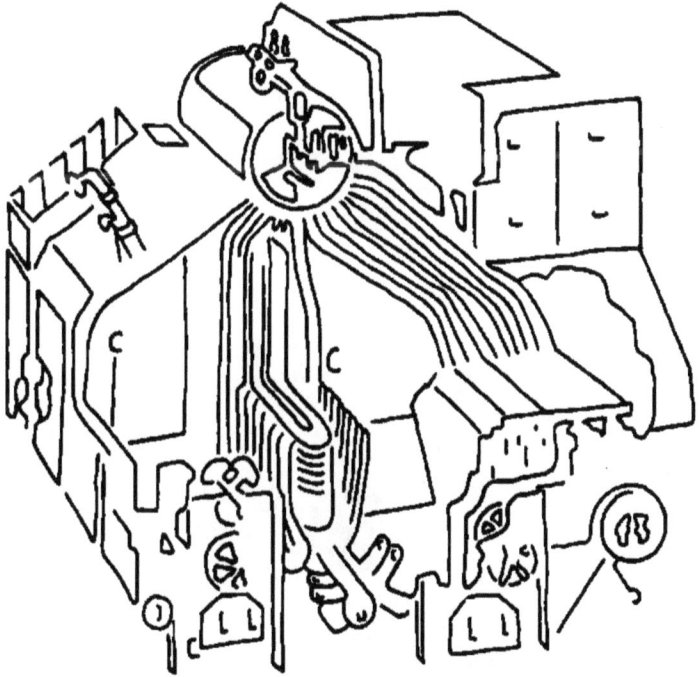

The cultural phenomenon that occurred following World War II was characterized by an almost insatiable demand for consumer goods and a striving to raise the standard of living. These twin goals greatly stimulated the growth of technologies that could help fulfill the needs and desires of Americans. Thus the period from 1958 through 1967 was to be the decade of nuclear development, electronics, computers, and exploration in space. The magnetism of these new technologies heralded the beginning of a new era. It was estimated that, by the year 2000, annual electricity requirements in the United States would be 8 trillion kilowatt hours, more than seven times their 1966 level of 1,140 billion kilowatt hours.

Looking to this future, the power industry began to expand facilities and make long-range plans for accommodating the burgeoning needs of the public, the government, industry, and the business world. The trend in research and development was ever upward, with increasing expenditures allocated to R&D and related activities of all kinds. It was not unusual for industrial manufacturers in this field to allocate as much as 25 percent of their operating income for research. As in the recent past, they set their sights on boilers, pressure vessels, and related components to function under greater and greater pressures and, when subjected, to higher and higher temperatures.

Bigness was "in." In 1967, for example, the Tennessee Valley Authority set a record by installing the world's largest pressure boiler, designed to deliver more than nine million pounds of steam at a pressure of 3,650 psi to generate 1,300,000 kilowatts of electricity.

During the 1950s, the record showed an unwavering and continuing increase in the demand for electricity by private consumers, industry, and government at all levels. To fill this need, utilities required larger and more efficient steam generating units. Fossil-fueled at first, they gradually made room for newer types that would use nuclear energy instead of coal or petroleum. Nuclear power seemed to be the answer to the problem of trying to satisfy the intense "hunger" of conventional boilers, some of which consumed more than 10,000 tons of coal a day. Later, the problem was to be intensified with the era of oil shortages and the demands of the OPEC nations that controlled large proportions of the world's oil and natural gas.

The economics of atomic-energy power systems — even without the fear of shortages of conventional fuels — seemed to justify the enormous expenditures required to plan and construct nuclear power plants, long before they could ever go on stream. Of major importance was the training of engineers, technologists, and the highly skilled personnel needed to operate nuclear plants, especially in light of the public's growing concern about safety and the threats of radiation and contamination.

181

Both expensive and time consuming was the unfamiliar work on the part of designers, manufacturers, and installers to obtain construction and licensing permits, not to mention compliance with the greatly expanded and far more complex rules and standards of the Code. Despite the fact that many components, such as pressure vessels, were enormous by comparison with their counterparts of a decade or two earlier, they had to be fabricated and machined to very precise tolerances. To accomplish the advanced production techniques required called for extremely large machine tools and the skilled manpower needed to operate them.

Advanced fabricating facilities had the capabilities, by the mid-1960s, to produce pressure vessels, for example that were 30 feet in diameter, more than 100 feet long, and weighed as much as 1,000 tons — yet as precise as the tools of a clockmaker when it came to staying within prescribed tolerances.

A great deal of engineering expertise was expended, as it had been during the previous decade, in the recovery of waste heat and the utilization of byproduct fuels that had previously been discarded or, at best, used inefficiently and often at random. New types of steam generators and accessories for existing types offered ingenious methods for accomplishing these objectives, reducing operating costs and often solving air pollution problems as well.

"What lies ahead?"

This question was asked and answered in an ASME paper, "Pressure Vessels and Piping: Design and Analysis," published in 1972 and characterizing the state of the art at that time. "One of the more significant developments at the present is the preparation of new design criteria for use in the temperature range above 800°F. . . . High-temperature design criteria under development are expected to follow the 'design by analysis' philosophy, namely consideration of all possible modes of failure, and design criteria associated with each mode.

"It is almost certain that in the next decade design methods and criteria will advance even more than in the last. Even at this writing, general computer programs to analyze cyclic, nonlinear, time-dependent deformations are available. The development of rational design criteria against which to evaluate local and general creep and stress relaxation, fatigue in the creep range, and other time-dependent structural behavior is now under way. As criteria are developed, structures designed and built, and experience gained, pressures for even further advances will result from the lessons learned."

Based on earlier research and testing, the subject of graphitization was still constantly in the minds of subcommittee members working on codes and specifications. Much of this was based on a summary report from the joint EEI-AEIC study in the late 1940s, Investigation of Graphitization of Piping.

This followed up an innovative research program undertaken at the Battelle Memorial Institute. As the report explained, "The failure in January, 1943, of a welded joint in a high-pressure steam line at the Springdale Station of the West Penn Power Company made the whole power industry, as well as the metallurgical world, acutely aware, *for the first time,* that graphite could form in steel piping operating at steam temperatures, and that its presence in certain instances could so weaken the piping as to produce a service hazard."

As a result of this, a number of investigations were undertaken, at Battelle and elsewhere, to study the fundamental causes of graphitization and the restoration of graphitized joints. The summary included reports from some 40 operating companies on the condition of their steam lines. Although the investigating committee came up with a few answers, it had to admit to a "tentative position" because this was an entirely new field of research and one whose questions had not been previously anticipated.

By the end of the decade, it was reported that plain carbon steel had been observed to be more susceptible to graphitization than carbon-molybdenum steel because of the general effect of molybdenum as a mild inhibitor. Silicon-killed steels, however, were shown to be "completely immune to graphitization," at least for the lengths of time during which the samples had been exposed — almost 500 days at temperatures up to 1,125°F. It was further discovered that vanadium and titanium acted as effective graphitization inhibitors when added to various steel alloys.

Research in the field of postweld heat treatment was also significant at the beginning of the 1950s, demonstrating that piping that had been thus treated to relieve stresses was less susceptible to graphitization. This was particularly true when temperatures higher than the then conventional 1,150°F to 1,250°F were used for the postweld heat treating. A temperature of 1,400°F was settled upon as the most promising for the prevention of this problem.

During these years, it had been the observation of engineers at Babcock & Wilcox, which could claim about as much experience with the subject as any firm in the industry, that carbon steel had a tendency to graphitization in the region of the welds at temperatures over 800°F. Consequently, carbon-molybdenum steel had been commonly used for piping in central service stations. Long-time service, however, had demonstrated that piping of this material was also graphitizing "with a chain-like formation" in the heat affected areas of welds and random formations in the body of the pipe proper. This common phenomenon had made it necessary to reweld and replace a considerable amount of piping in the power industry. It was learned, too, that the rate and extent of graphitization in piping materials had been affected to some degree by steel manufacturing practices.

In an earlier report on the subject, a subcommittee of the ASME Boiler and Pressure Vessel Code had made some similar observations on piping at elevated temperatures and had come to the conclusion that "the art is yet in its developmental stage."

By this time, numerous alloys had been developed for high-temperature service, which then was in the range of 750°F to 1,100°F. Most of these were covered in the standards and specifications for the American Society for Testing and Materials. However, the alloy which was receiving the greatest amount of favorable attention, and most markedly in the piping field, was carbon-molybdenum steel. The committee reported, as B&W was to find itself in agreement with, that this steel "was being used almost exclusively for high-temperature, high-pressure applications, although it was by no means the only alloy readily available for the purpose." Although it did not have as good high-temperature characteristics as several other alloys, it was regarded as the most economical, was dependable, and was "the most nearly foolproof of all the alloys available.

One of the most pressing needs of the day was a simple test that could be applied by users to examine the properties of carbon-molybdenum steel. The standard creep test was too long, requiring a minimum of 1,000 hours to complete. Other tests were fallible because they introduced uncontrolled variants. The net result was that, in the end, users had to subject the alloys to a variety of tests, such as time-yield, varying rate tensile, stress-rupture, and others, in order to be certain about the properties of the metal under examination. This was a telling example of a research problem that was ever-present in the industry and with which the Code Committee was constantly wrestling: simplifying the methods and systems for examining materials and components. Coordination and communication were two of the most vital factors in the procedures followed by ASME committees and those of other societies and associations in order to benefit from each other's research, and to avoid as much as possible unnecessary duplication of effort. For example, the Welding Research Council was constantly engaged in studies and investigations along the lines of those mentioned above and at all times made its efforts known to the other groups.

In 1962, the Executive Committee of the Welding Research Council appointed a Program Evaluation Committee to summarize and present the results of research by the Pressure Vessel Research Committee in a more useful form for designers, the ASME Code, and other codes and standards. One example was the work of the Materials Division, which undertook in the mid-1960s to determine the effect of metallurgical treatments and geometrical considerations on the suitability of steels for pressure vessel service. These treatments included aging, straining, normalizing, quenching, and tempering. Geometrical considerations included thickness and discontinuity effects.

Instituting a new program at Lehigh University in 1968, a special Task Group undertook a number of significant studies that included the influence of warm prestressing on the strength, ductility, and toughness of pressure vessel steels; a study of slow cycle rates on the fatigue strength of pressure-vessel steels at elevated temperatures; and compiling data on the mechanical properties of heavy-section steel plates, weldments, forgings, and castings and determining the effectiveness of nondestructive examination methods.

"The WRC was always very careful in its efforts to make sure that we at the ASME were aware of its plans for research and its ongoing investigations," explained retired engineer Frank Williams, a member of many Code committees and something of a student of Code history. "The same was true of just about every other professional society I knew of. In fact, if you will read the introductions to their articles of incorporation or similar documents, you will find a clause to the effect that one basic objective is to communicate with other professional bodies in their field, exchange information and data, and avoid costly duplication of effort. The Materials Properties Council is another good example of this kind of coordination."

The Materials Properties Council had been established in 1966 with the following objectives: to compile reliable data on the engineering properties of metals and alloys; to evaluate such data as might be useful; to make new facts quickly available through reports, correspondence, and publications; to keep informed about research in the materials field to avoid any duplication of effort; and to act as co-sponsor with ASME and ASTM of research on the effects of temperature on the properties of metals.

Since its founding, the MPC has continuously provided vehicles within which producers, fabricators, users, and others could work together on engineering problems in which the basic characteristics of materials represented key factors in the determination of codes and standards.

"The field of materials properties is a bewildering area of conflicts, difficult comparisons, and fragmentary knowledge," stated the MPC as recently as 1987, despite the fact that the Council has always focused its attention on "trying to do *one* thing well," providing engineering data on materials used in the industry.

During its first decade, MPC was distinguishing itself by undertaking projects that ranged from boiler tubes to weldments, from bolting to turbines, investigating fatigue, stress-rupture, fracture mechanics, tensile strength, creep, and corrosion, among other conditions. The organization's track record of accomplishment was excellent because, like the various code committees, it had generated a membership of professionals who were experienced and

185

thoroughly qualified, often among the top specialists in their fields of endeavor.

Typical assignments included stress-rupture studies of low-strength carbon steel, the investigation of tensile strengths in selected alloys, studies of weldment creep, tests to determine the nature and extent of hydrogen embrittlement, high- and low-cycle fatigue testing of boiler-code materials, an examination into corrosion of heat-exchanger components, and the assessment of dissimilar metal welds in boilers.

"We regularly analyze and evaluate property data which ASME needs for Codes and Standards work," explains a current synopsis of MPC's policies and activities.* "We are involved in programs to coordinate the utility industry's steam piping and turbine rotor evaluations. We worked with the Atomic Industrial Forum (AIF) to prepare a comprehensive program to prevent bolting problems in nuclear power plants and have studied the effects of radiation on transition temperatures of ferritic materials."

Some of the organizations supporting the Materials Properties Council, in addition to ASME, have been the American Petroleum Institute, American Society for Metals, American Welding Society, Electric Power Research Institute, American Society for Testing and Materials, and the National Bureau of Standards.

Another organization strongly involved in this field, but which came into the picture later was the Federation of Materials Societies. FMS was founded in 1972 as "a body of technical and professional societies having dominant capabilities in materials science and engineering. Its purpose is the advancement of materials science and materials engineering, and the advancement of their interaction with other science and technologies and with public and private institutions through cooperative efforts."

Several key factors motivated the formation of FMS: the recognition of *materials* as being of vital importance in human affairs; the fact that one-third of all physical scientists and engineers are involved in material science and engineering; and a situation that found itself with some 35 technical societies having a recognized interest in materials, yet with a fragmentation of this enormous resource that was inhibiting unified competence regarding fundamental materials problems.

The formation of the FMS enhanced the capabilities of individuals and organizations in the industry to identify common problems, stimulate more collaboration on projects of mutual interest, improve the knowledge and education of those who were dealing with materials, improve public understanding of the importance of materials to society and the need for better conservation, and provide a mechanism for effectively communicating with

---

* MFC member bodies are ASME, ASTM, ASH, and AWS.

government bodies and thus stimulating legislation favorable to this special field.

The Federation of Materials Societies also became active in promoting a National Materials Policy in an effort to help safeguard American industry, jobs, and security. In the course of developing its major power resources, the United States had long since become the top nation in providing electricity and almost every other form of energy. At the same time, however, it had become dependent upon other nations to supply many of the critical materials required for engineering projects of all kinds. Shortly after its founding, the FMS estimated that "the United States depends upon foreign imports of 50 to 100 percent of our total requirements for many metals and minerals which are vital to our industrial and defense needs."

Many deposits of urgently needed metals such as cobalt, chromium, columbium, manganese, and the platinum family were located in very remote regions or in areas marked by social instability, political strife, or active warfare. Situations like these, when combined with competing demands for minerals and the spread of local hostilities, created sharp fluctuations in the availability and prices of the materials. Shortages of supply, exorbitant prices, and related factors often had direct bearings on codes and standards since specifications had to make allowances in some cases for substitute metals and minerals whose properties and characteristics were inferior to the ones preferred.

The sponsoring membership of the Federation of Material Societies was composed not only of ASME and the Institute of Electrical and Electronic Engineers, but of a number of associations directly involved with materials, such as the American Society for Metals, the National Association of Corrosion Engineers, the American Society for Testing and Materials, and the Society for the Advancement of Material and Process Engineering. Thus it was not surprising that one of the major goals was to take measures to assure the availability of critical metals and minerals by stockpiling them, encouraging exploration in North America for domestic resources, and undertaking research to perfect substitutes that could be used "without degradation in the performance of manufactured products."

Conservation was also a major issue, aimed at promoting energy conservation in materials extraction, production, and processing, as well as the increased recovery and recycling of materials in municipal and industrial wastes and in products that were unserviceable. Hand in hand with this was the effort to increase the durability of products and reduce the rate of replacements by the improved application of materials to avoid premature failures caused by fracture, wear, and corrosion.

Starting in the 1950s, right after World War II, the dependence of the United States on foreign sources started to increase markedly, in part because of the increasing use of new kinds of alloys whose properties were especially favorable to the development of boilers, pressure vessels, and other products and components used in the power and energy industries. Foreign sources, for example, supplied more than 90 percent of America's total annual requirements for columbium, manganese, chromium, cobalt, bauxite, and platinum metals; between 75 and 90 percent of the needs for tin and nickel; and from 50 to 75 percent of those for zinc, antimony, tungsten, and cadmium.

Manganese had come into great demand for the production of iron and steel, chromium for the production of stainless steels, and chromium, cobalt, columbium, and nickel for high temperature alloys — all of them high-priority ingredients for many of the materials and products that were directly covered by the ASME Code.

The question of *standards* has always been uppermost in the minds of the Boiler and Pressure Vessel Code Committee, a subject that was one of the basic motivating forces behind the formation of ASME itself. This viewpoint has also been in focus in just about every other professional engineering society during the past century in America. A significant and historic step was taken when, in 1918, five engineering societies and two federal agencies joined in the organization of the American Engineering Standards Council, the fundamental purpose being "to coordinate standards activities across interfaces, transition points, or discipline interfaces."

This was the beginning of a program that was to be of great significance, both historically and functionally, to the ASME Boiler and Pressure Vessel Code and its related operations. Over the years, the organization changed names several times, finally evolving, in 1969, into the American National Standards Institute, familiarly known as ANSI. With a federated membership that included about 180 organizations in virtually every technical discipline, ANSI became the national coordinator of voluntary standards development, as well as a clearinghouse for information on standards, foreign and domestic alike.

The Institute's approval procedures for prospective standards required a consensus of interest by all parties that might be affected by each proposal. Its requirement for due process maintained confidence in, and credibility for, American National Standards. The Committee Method was eliminated, but the Canvass Method was retained, by which a proposed standard could be submitted to ANSI, along with a canvass list. Those on the canvass list had the opportunity to review, comment, and vote on each proposed standard.

As the national coordinator of voluntary standards, ANSI was to become involved in programs that were new and in evolutionary stages, as well as in ones where there might be conflicting standards or proposals. For example, when the U.S. Geological Survey recognized a demand for certification and standards for offshore oil and gas operations, it advised ANSI of the need. After a thorough study and review of the situation, ANSI then requested ASME to define the necessary procedures for administering a certification and laboratory accreditation program. ASME agreed to undertake the project, estimating that it would be able to develop the standards and procedures within one year (a short period of time for an action of such complexity and breadth) because of its past experience with similar functions. As it turned out, from the time of the first committee meetings, through the hearings, approvals, and issuance of certificates, the elapsed time was just about 12 months. That period included the necessary surveys and the issuance of certificates for manufacturers, assemblers, and laboratories involved in the concerted offshore operations.

"We are always on the threshold of new challenges," wrote the late Dr. David S. Jacobus. "No sooner do we become satisfied that we have taken all the latest technological advances into account than something completely new and unexpected shows up at the back door."

One of the major challenges that fit this comment by Dr. Jacobus was certainly the field of nuclear energy, bringing about massive efforts on the part of engineers and other professionals to pool their knowledge, exchange ideas, and try, through joint efforts, to anticipate the problems and demands of the future. The mid-1950s, for example, saw the formation of a Nuclear Congress General Committee, the basic objectives of which were to achieve maximum unity among the various engineering and scientific societies concerned with nuclear energy and related problems; to foster a free exchange of knowledge, information, and ideas between engineers and scientists in all fields of technology; and to provide an organizational framework for mutual participation and planning in the nuclear field.

The American Nuclear Society was appointed to serve as the general coordinator for annual Congresses, coordinating its activities with the Engineers Joint Council. The plan of action called for conferences to determine the progress and state of nuclear science and its peacetime applications and to establish policies. The Nuclear Congress did not investigate specific developments, such as the design and planning of nuclear power plants, but rather limited itself to high-level considerations. One of its most practical efforts was directed at keeping the public, as well as the technical world, aware of nuclear developments and promoting nuclear energy "For Mankind's Progress."

On the firing line, organizations like ASME became charged with the task of handling the details, as for example the specifics relating to reactor construction and operation. In July 1968, the requirements for a quality assurance program became mandatory for nuclear certification. ASME now required a nuclear survey team to evaluate the implementation of all quality assurance programs to make certain that they satisfied the intent of the Code. Each survey team consisted of an ASME team leader, an ASME consultant, a consultant from the National Board, an authorized inspection agency representative, and a state or provincial representative. Prior to that date, the issuance of a certificate had been based upon reports from the authorized inspection agency and the state (or provincial) jurisdictional authority.

Starting in the late 1960s, Section III expanded from pressure vessels to include piping, pumps, valves, concrete reactors and containments, core supports, and component supports. The scope of the nuclear certification program grew right along with the scope of Section III.

By the fall of 1972, ASME certification was extended from the United States and Canada to the rest of the world, thus becoming, in reality, an international certification program.

CHAPTER 17

ACHIEVEMENTS AND SETBACKS

Forces of Change, Development, and Opposition

From its very inception in 1914, the ASME Boiler Code Committee was to feel the bitter lash of controversy, made all the more biting in light of the fact that many individuals had volunteered long hours of work, with little or no recompense, to establish an institution that would benefit not only the engineering profession but all of society.

The first such smudge on the history of the Code was known as "The Parker Case," in reference to one John Clinton Parker who slammed into a personal confrontation with the Committee, even before the Code came off the press. In 1914, when the Code was still in draft, Parker was in the business of manufacturing steam boilers in Philadelphia. As a confirmed individualist, he resented the fact that any outsiders — most particularly those damned New Yorkers — were going to dictate to him how he should build his products. He summarily ripped off a letter of protest to the ASME Council.

"The writer desires to register a strong protest against further backing of the propaganda for state control of boiler design, with the funds, and at the meetings and in the publications of this Society."

As Bruce Sinclair explained the action in his classic history of ASME, "Parker saw the work of the Boiler Code Committee as the devious work of special interests to sabotage their competition and he darkly hinted at evidence demonstrating 'that these interests do not stop at underhanded means to accomplish their ends.' In a way that would become all too familiar, Parker's attacks on the Boiler Code were full of personal slander and extravagant language."

Parker was so determined and unflagging in his efforts to force the Code into oblivion that the litigation he fomented — with court investigations, countersuits, and appeals — lasted from 1915 until well into the 1930's. He also harassed ASME members and those on various Code committees with printed literature intended to undermine the credibility of the rules and standards and create unfavorable publicity on all fronts. He finally climaxed his obsessive attacks in 1933 by alleging that the secretary of ASME and his colleagues were mismanaging Society finances to the tune of a quarter of a million dollars.

Ironically, in the end Parker's vituperative campaign backfired and motivated beneficial actions that might never have been taken otherwise. Realizing how many members, not to mention the general public, had been misinformed and misled by The Parker Case, ASME began to establish programs to enhance communications, both internal and external, to establish a better dialogue all around. This was to pay off in the future in many ways.

For some 75 years, in the presentation of information relating to codes, standards, and accreditation programs, ASME has included in its bylaws, procedures, and policies provisions to provide for "due process of law," to protect the rights of all individuals or groups with interest in the subjects of concern. ASME recognized that its programs were used as a means of satisfying regulatory requirements and contractual agreements and that the public needed to be party to the legal development of codes and standards.

Over the years, there have been legal actions that have caused the ASME to revise its bylaws, procedures, and policies in an effort to avoid legal adversities. Some of these have directly involved the nature and administration of codes, standards, and accreditation programs.

Until the early 1960s, there was little apparent interest by the federal government in these codes, standards, or accreditation policies. Although American states and cities and the provinces of Canada had long been referencing ASME codes as a means of satisfying government regulatory requirements, the United States government had remained silent. Its lack of attention was particularly evident in the matter of the Boiler and Pressure Vessel Code.

All this changed abruptly, however, during the Kennedy Administration when a foreign delegate charged openly that ASME accreditation was a non-tariff barrier to trade. At that time, ASME only gave accreditation to manufacturers located in the United States or Canada. There was a valid reason behind this geographical limitation: the infrastructure of the accreditation program consisted of ASME, insurance companies, and the states or provinces concerned. Applicants, who needed no formal review for accreditation (then known as certification) had merely to indicate that they had arranged for authorized inspection services. With confirmation of this fact from either an insurance company or area jurisdiction, ASME then granted the accreditation.

The manufacturer was thereby authorized to affix the ASME Code Symbol on products that had been designed, fabricated, and inspected in accordance with Code requirements. In July 1968, an event occurred that prompted government action. At that time, ASME announced that manufacturers of nuclear components were required to have a quality assurance program that would be surveyed by an ASME team and accepted by ASME through committee procedures. This change caused ASME, the National Board, inspection agencies, and local jurisdictions to become involved jointly in all surveys of applicants for nuclear accreditation.

This development placed a new perspective on ASME accreditation in a number of ways. For one thing, the federal government now saw a way for ASME to expand its geographical scope of operations. In 1968, the Society was served with a civil investigative demand and in July 1970, was sued, along with the National Board, by the Justice Department, under the Sherman Antitrust Act.

For the next two years, ASME and the Board negotiated with the government and in September 1972 signed a consent degree, which provided for federal overview and implementation. From that period on, both organizations were recognized as being fully international in scope. The decree provided that ASME could require all manufacturers located outside of the United States and Canada to register with the National Board those products bearing the ASME Code Symbol. This procedure aided the Committee in tracing products that subsequently were found to deviate from Code requirements. Furthermore, the decree cited the United States as a host government to aid in the overseeing of these global accreditation programs.

Since it was now required that foreign manufacturers register boilers and pressure vessels with the National Board, the Board began to serve as a leader of review teams for those applicants located outside the United States and Canada. In addition, the National Board provided team leaders for reviews in cases in which a state or province was the inspection agency and had requested the Board to serve as its agent.

As has been mentioned earlier, since the first Code was published in 1915, the ASME Boiler and Pressure Vessel Committee has answered inquiries about Code requirements, often reaching interpretations that have brought about revisions or additions. The number of inquiries, few at the beginning, mounted by the hundreds and thousands as the years went by. In 1970, there were about 14,000 such inquiries — some fairly simple and others very complex. Among these was an inquiry about a low-water fuel cut-off device that involved a time delay.

As was the case, the standard procedures for responding were followed and an interpretation provided. Sometime later, it was alleged by a newspaper that the business of a small local manufacturer had been seriously damaged by a competitor's use of this Code interpretation. A Senate investigation then resulted from the newspaper's charges, followed by a lawsuit alleging violations by ASME of antitrust and monopoly laws. This lawsuit became known as the "Hydrolevel Case," referring to the name of the manufacturer involved, and was heard by the Supreme Court of the United States, as well as by subordinate courts.

During the district court trial and court of appeals proceedings, it was brought out that ASME policy on handling interpretations provides for the publication of all interpretations. The intent was to enhance Code user awareness of interpretations and thus avoid situations in which one organization might unknowingly damage the business or reputation of another. The Supreme Court expressly commended ASME for recognizing its responsibility to communicate more effectively in this way with all those who used the Code or were affected by it.

The Hydrolevel case also highlighted the Society's acceptance of responsibility for the actions of all volunteers serving in its name on codes, standards, and accreditation committees. The Second Circuit Court of Appeals stated the opinion that volunteers had the "apparent authority to act in behalf of ASME," a viewpoint that was upheld by the Supreme Court. What it meant, in effect, was that the interpretations of inquiries by volunteer committees were just as "official" as they would have been if stated by salaried employees and officers of the Society.

Because of the provisions in the bylaws for the indemnification of an individual while serving on an ASME committee, and because of the Court's consideration of an organization's responsibilities for its volunteers, ASME revised its policies on conflicts of interest and the Code of Ethics of Engineers. In the winter of 1982, the Board of Governors instituted the following requirement: "Each member of the Council, Board, Committee, Subcommittee, or other decision-making body and each individual elected or appointed to act for, or on behalf of, ASME shall indicate in writing that he or she accepts the conditions of this Society policy."

To implement this policy, the Council on Codes and Standards made it mandatory that all persons appointed to these positions would be given copies of the Code of Ethics of Engineers and the statement of policy on conflict of interest, along with a letter of transmittal from the Council. The letter stated that the recipient, by accepting the position and signing the letter of transmittal, affirmed a willingness and responsibility to follow the Code of Ethics and ASME policy on conflict of interest.

In addition to receiving these documents, all staff members and volunteers associated with codes, standards, and accreditation were provided with supporting guides on compliance with antitrust laws and tort laws. It was explained to them that the antitrust guide had been adopted by the Council in order to ensure adherence to the "due process" requirements in the development of codes and standards and the administration of accreditation. Further, it was pointed out that the tort — or product liability — guide was included as a pertinent explanation of a legal area that applied to antitrust-law compliance during the administration of codes, standards, and accreditation actions.

Speaking out at the time the Hydrolevel case was still a matter of controversy, Melvin R. Green, Managing Director of ASME's Codes and Standards Department, reminded the engineering profession that, "although ASME develops more than 400 codes and standards and has reduced the number of boiler accidents through such action, the public does not appreciate the fact that our principal purpose for developing codes and standards is public health, safety, and welfare. Because of this lack of understanding, the legislative and executive branches have been able to continue in their attempts to regulate voluntary standards and certification activities in the United States."

It was pointed out at a National Board conference in the spring of 1988 that ASME had testified in 1975 against federal regulation, in 1976 against the Voluntary Standards and Certification Act, in 1977 against another part of the same title, and in 1978 had numerous interactions with the Office of Management and Budget of the United States, which resulted in an OMB circular that the Society supported. The subject covered federal agency interactions with voluntary standards organizations.

Not satisfied with the extent of the OMB circular, the Federal Trade Commission proposed rules for voluntary standards and certification, published in the *Federal Register*. ASME provided comments about two principal subject areas: (1) that the proposed rules were in violation of the First Amendment of the Constitution because of their censorship provisions; (2) that the FTC did not have any jurisdiction over a "learned society" as defined in this instance.

To those who questioned whether there was any hope for voluntary standards and certification activities, in light of the federal government's efforts to undermine them, the ASME Code Committee emphasized positive examples of accomplishments during the 1970s. This "inventory" of the results of ten years of federal interaction included the following examples.

- ASME and the National Board extended certification and registration from the United States and Canada to the rest of the world. This international system of certification was made possible by the persistence of the two bodies in their efforts to provide not only for safety but for procedural "due process."
- As a result of public clamor stimulated by ASME, the proposed Voluntary Standards and Certification Act did not become law. The key to this defeat was the work on the part of organizations such as ASME and the National Board to inform the public that ANSI was able to set the criteria and enforce the rules through a more democratic and cost-effective means than the federal government would be able to accomplish. The response of an informed public was seen in the defeat of the measure.
- The OMB circular was revised because of public comments to the extent that it could provide uniform criteria for federal participation in voluntary standards activities, leaving no question among federal representatives as to what the public policy would be.

- More federal agencies than ever before began themselves to make use of voluntary standards.
- Many American standards became de facto international standards. It was further pointed out that states, municipalities, and provincial governments considered safety, public welfare, and health to be essentials that lay within their areas of responsibility and authority. Hence, they vigorously supported systems with which they could *interact* in the matter of voluntary codes and standards and not ones that were prescribed by Washington, D.C.

As ASME expressed it at the time, "When the public recognizes the benefits of the voluntary standards system, this system will be the 'wave of the future,' with federal *participation,* yet without federal *domination.*"

The event that made the greatest impact on the ASME Boiler and Pressure Vessel Code — from the public viewpoint at least — was the aforementioned Hydrolevel case. The essence of the matter was summed up in a well-balanced manner in the June 1975 issue of *Mechanical Engineering,* before the courts had reached a final decision.

"The setting is the U.S. Senate. The action — hearings before the Subcommittee on Antitrust and Monopoly of the Senate Committee on the Judiciary. The question — did ASME Code developers use improper influence in an attempt to discredit the product of a small manufacturer?"

As reviewed during Senate hearings in March 1975 explained the editorial, the substance and chronology of events surrounding this question were as follows: McDonnell & Miller, Inc., which then marketed some 85 percent of the heating boiler safety controls in the United States, followed ASME's longtime procedure when engineering questions arose and submitted an inquiry to the Boiler Code Committee. The question, as phrased by the company was as follows.

"Par. HG-605 of Section IV of the Code states that each automatically fired steam or vapor-system boiler shall have an automatic low-water fuel cut-off so located as to automatically cut off the fuel supply when the surface of the water falls to the lowest visible part of the water-gauge glass. Is it the intent of this statement that the cut-off operate immediately when the boiler water level falls to the lowest visible part of the water glass, or is it permissable to incorporate a time-delay feature in the cutoff so that it will operate after the boiler water reaches some point below the visible range of the gauge glass?"

Was this a straightforward question? Or was it — as another manufacturer of similar boiler cut-off devices, the Hydrolevel Corporation, objected — intended to impugn the worth and safety of the company's product?

Hydrolevel objected too that ASME's answer to this inquiry was equally damaging, at the same time inferring that McDonnell & Miller had used the influence of its vice president of research, who was also vice chairman of the Society's Subcommittee on Heating Boilers, to ensure that the interpretation defined would be unfavorable to Hydrolevel.

The McDonnell & Miller letter of inquiry followed the same normal procedure as the thousands of others that had preceded it in previous months and years. It was directed to the staff secretary of the Boiler and Pressure Vessel Committee, who in turn placed it in the hands of the proper subcommittee chairman for action. The chairman's response was, in due course of time, reviewed by the chairman and vice chairman of the Boiler and Pressure Vessel Committee. The substance of this response was recorded as follows.

"A low-water fuel cut-off is considered strictly as a safety device and not as some kind of an operating control. Assuming that the water gauge glass is located in accordance with the requirements of Par. HG-602(b), it is the intent of Par. HG-605 (a) that the low-water fuel cut-off operate immediately and positively when the boiler water level falls to the lowest visible part of the water gauge glass.

"There are many and varied designs of heating boilers. If a time delay feature were incorporated in a low-water fuel cut-off, there would be no positive assurance that the boiler water level would not fall to a dangerous point during a time delay period."

The Hydrolevel complaint asserted that the McDonnell & Miller company then deliberately gave copies of this letter to its salespeople for use in their calls on customers. Applied in this manner, charged Hydrolevel, the interpretation slyly suggested that Hydrolevel's automatic low-water fuel cut-off devices, which did not have or require a time-delay feature, were unsafe and did not meet Code requirements.

A vice president of McDonnell & Miller protested that the letter had not been forwarded to his sales offices as a selling tool and that only three salesmen had actually even seen it. This explanation did not satisfy Hydrolevel, which sent an incensed protest to ASME, a letter that was quickly turned over to the full Subcommittee on Heating Boilers. ASME then made it clear that there was no intent to prohibit the use of low-water cut-offs having time delays and that such devices could meet Code requirements as long as they were properly installed. ASME's response further emphasized the fact that the Society did not then, and never had in the past, approved or disapproved the quality or reliability of any specific company's products — nor did it have control over the use or misinterpretation of this or any other information it might issue.

199

It was thought by most of the people involved in, or knowledgeable about, the issue at hand that ASME's response sent to Hydrolevel on June 9, 1972, had cleared the atmosphere and left the matter resolved. However — more than two years later, on July 9, 1974 — *The Wall Street Journal* ran an article on the controversy. The editorial criticized both ASME and McDonnell & Miller for what the editors asserted was an unacceptable handling of the situation. The article condemned the original letter of interpretation from ASME to McDonnell as an instrument to "spread the rumor" that the Hydrolevel product failed to meet Code requirements. Besides, charged the *Journal*, "the Hydrolevel dispute raises serious questions about the close ties between the dominant company in an industry and the professional society that serves as its watchdog."

"Publication of this adverse piece in such a well respected newspaper," reported *Mechanical Engineering* in its issue of June 1975, "caused quite a stir. In response to the expressed concern of various Society officers, ASME's Professional Practice Committee took up the matter. They investigated the question of whether there was any evidence of unethical conduct on the part of any ASME member. The Committee first discussed the case at a September 1974, meeting. They promulgated a resolution exonerating James Airman of the subcommittee that had drafted the interpretation at the 1974 Winter Annual Meeting, and the resolution was finalized on January 20, 1975."

"In its own effort to resolve the problem," wrote Bruce Sinclair in 1980 in his history of ASME, "ASME's Professional Practice Committee operated without the crucial information that subcommittee personnel had drafted the original letter of inquiry as well as the responses to it." This lack of information weakened ASME's stand.

From that point, the controversy became the subject of a Senate hearing before the subcommittee on antitrust and monopoly.

The situation was muddied when, as another editorial in *Mechanical Engineering* phrased it, *"The Wall Street Journal* struck again. This time, even with all the facts out in the open, they managed to include serious inaccuracies in their report. . . ."

This second article centered on the premises that the Senate hearings had proved that ASME committee members were guilty of improper actions in the drafting of letters and interpretations. The article further suggested that McDonnell & Miller had admitted that it intentionally misinterpreted ASME's response, to the disadvantage of Hydrolevel. "In fact, however," *explained Mechanical Engineering,* "the Senate subcommittee has come to neither of these conclusions, nor has it issued a statement to suggest they found evidence of wrongdoing."

Not content with these charges, the *Journal* article portrayed ASME as having been *duped* by the McDonnell company in the first place, in an effort to discredit the products of its competitor, Hydrolevel. Although the President of ASME wrote a letter to the editor of the *Journal,* citing the errors and innuendoes in the article in its issue of April 7, 1975, the damage had been done.

What followed was the kind of lawsuit that the Society had sought to avoid from the very beginnings of the Boiler and Pressure Vessel Code by establishing procedures that nurtured fairness and objectivity. Critics of industry codes and standards had long predicted that this kind of confrontation was inevitable. Now they would have their day in court.

The lawsuit was long and exhausting. Two other defendants in the case, International Telephone & Telegraph (parent company of McDonnell) and Hartford Steam Boiler Inspection and Insurance Company, settled out of court in 1978. But ASME, believing steadfastly that it was not guilty of antitrust or monopoly actions, continued the fight. For its efforts, it was assessed damages of $3.3 million, trebled by federal law to $9.9 million when a United States District Court ruled in 1979 that the Society was guilty of violating antitrust statutes. Aghast at the enormity of the assessment, and even more disturbed by the findings that ASME had been labelled a conspirator, the Society appealed the decision, but to no avail.

Historically — apart from the great cost, the demands on the time of members who would otherwise be engaged in productive work, and the tarnish on the ASME image — the most significant effect of the lawsuit was that it weakened the whole structure and philosophy of codes and standards that had been evolving for 65 years. It provided the federal government with a wedge so that it could pry its way more firmly into the regulation of codes and standards activities in the United States. The system of *voluntary* standards, which ASME and the states, municipalities, and provinces had viewed as "the wave of the future," were now at stake.

Looking at the case in hindsight, it was seen that it was not so much the actuality of any violation as it was the *appearance* of conflict of interest that was the most damaging to voluntary codes and standards. Sinclair's history commended Melvin R. Green for devoting much of his time and energies during this critical period defending the voluntary standards system at congressional hearings and before other groups involved. Green, wrote Sinclair, explained that the procedures sanctioned by the American National Standards Institute had been followed with integrity in the Hydrolevel case. "But even as the Society's leaders waited for the verdict on their appeal of the District Court's judgement, the fundamental issue in Green's mind was still whether the country was better served by government-imposed standards — a prospect that conjured up endless bureaucracies staffed by the technically inept — or a system of voluntary standards which in a 'more democratic and a more cost-effective' way protected the public interest."

Even the worst setbacks, however, are not without their positive sides. The Hydrolevel case served one important function, motivated by the fact that there were many *internal* rumblings about procedures during the hearings, the trial, and the aftermath. Members themselves questioned the sequence of events that had led to the catastrophe. As a result of these actions — not to mention the outcome of the case itself — ASME reviewed its methods and procedures, tightened its policies, and in general took steps to minimize the chances of facing similar legal challenges in the future.

Sensitive to the possibilities of misinterpretation and misuse, the Committee also reviewed its policies regarding the Code Symbol Stamp, now familiar not only in North America but around the world, and ways to enhance its meaning and protect its integrity.

"It is the aim of the Society," explained a Statement of Policy on the Use of Code Symbols in Advertising, "to maintain the standing of the Code Symbols for the benefit of the users, the enforcement jurisdictions, and the holders of the Symbols who comply with all requirements. Based on that objective, the following policy has been established on the usage in advertising of facsimiles of the symbols and reference to Code construction."

The statement went on to emphasize that the Society did not "approve," "certify," "endorse," or "rate" any product or construction and that advertisers must avoid any statements or inferences that would so indicate. The farthest any holder of the Code Symbol and a Certificate of Authorization could go in an advertisement or promotion was to state that its products "are built in accordance with the requirements of the ASME Boiler and Pressure Vessel Code," or "meet the requirements of the ASME Boiler and Pressure Vessel Code."

Although the symbol itself could be used only for stamping and nameplates, the regulations did permit the use of facsimiles "for the purpose of fostering the use of such construction" by an association or society, or by an authorized holder who could use it in advertising to show that clearly specified products would carry it. General usage in a print advertisement or commercial was not permitted unless *all* of an advertiser's products were constructed under ASME rules.

Policing the use of the Symbol abroad was not always easy, because of distances and language barriers. Yet the Committee persisted and very few violations were noted. The first mention of the issuance of a Code Symbol Stamp to a foreign manufacturer (other than North American) had appeared in the January 12, 1973, minutes of the Main Committee. This was a pressure vessel stamp issued to De Dietrich & Cie., 1 Rue d'Offwiller, Zinswiller, France. The next mention had been in the March 9, 1973, minutes, which showed the issuance of pressure vessel stamps to two British and two Japanese manufacturers. Subsequent minutes had recorded the issuance of stamps in various categories, including nuclear, to manufacturers in Great Britain, France, West Germany, Italy, Japan, and other countries.

Despite the exhaustive and exhausting Hydrolevel case, ASME did not break stride in its many operations, not even in those directly related to the Boiler and Pressure Vessel Code. One of the more positive developments in the era of the 1970s was the matter of looking backward.

"We can only marvel at the old-time achievements on realizing they were accomplished without the handbooks, theoretical foundations, and encyclopaedic knowledge at our disposal today. As a matter of fact, our funds of knowledge are the outgrowths of the seat-of-the-pants engineering we look back to."

This statement, in 1979, reflected the outlooks and attitudes of the History & Heritage Committee, established by ASME to acknowledge and preserve historic mechanical engineering landmarks. One of the Committee's goals when it was established in 1973 was described in this way: "to educate the general public, as well as engineers, about the world's rich technological heritage. The oral history, the heritage sites and collections, and the landmarks programs provide an excellent panorama of developments, from the ancient water wheel to the space station. Steam engines and iron works take us back to the nineteenth century, just as computers and automated production point out the relevance of mechanical engineering in our lives today."

The History and Heritage Committee was formed with three major programs in mind that related to mechanical engineering: (1) the reconstruction of landmarks, using real or restored artifacts that have contributed in an outstanding way to the development of humanity; (2) the establishment of museums and other collections that included objects of special significance in this field; and (3) the designation of sites, particular locales associated with some events, buildings, developments, or machines that were historically meaningful.

Many of these landmarks, objects, and sites related to those mechanical engineering areas associated directly with equipment and components covered under the Boiler and Pressure Vessel Code. Typical examples that can be seen and visited today include:

- Patriot's Point Naval and Maritime Museum, near Charleston, South Carolina, site of the N.S. *Savannah* in 1962, the first nuclear-powered cargo-passenger ship.
- The Edgar Steam-Electric Station in Weymouth, Massachusetts, whose high-pressure turbine and boiler set a new record for economy in the mid-1920s by producing electricity at the rate of one kilowatt hour per pound of coal at a time when it was not unusual to have to burn five to ten times that amount per kilowatt.
- Chattanooga, Tennessee, where the first fusion-welded steel drum was tested during 1930, leading to the commercial acceptance of welding for the fabrication of boiler drums.
- Shipping port, Pennsylvania, Atomic Power Station, the first commercial central electric-generating station in the United States to utilize nuclear energy, and which was dedicated by President Eisenhower on May 26, 1958.
- The Geysers Unit #1, in Sonoma County, north of San Francisco, which in 1960 went on stream as the first commercial geothermal electric-generating station in North America and which later became the largest geothermal complex in the world.
- Baldwin Articulated Steam Locomotive, in Sacramento, California, which operated between 1944 and 1956 and represented the final phase of steam locomotive development in both size and power.

The work of the History and Heritage Committee came to fruition in an ongoing program that escalated during the 1970s and has continued ever since.

Many of the other areas of accomplishment by ASME in general and the Boiler and Pressure Vessel Committee and subcommittees in particular during this era have already been recounted in these pages. But they were summed up in brief in a History of the 1965-1975 Decade of the ASTM-ASME-MPC Joint Committee on the occasion of its 50th Anniversary in June 1975.

"It is clear," said the Anniversary Report "that recent Committee activities have reflected, directly or indirectly, a number of important trends in our nation. Among these are the expansion of civilian application of nuclear power; the development of new power sources and attendant problems; rapid advances in the aerospace industry; the need for more precise design criteria; the need for more economical solutions to engineering problems; the growing awareness of a need for environmental protection; legal and governmental restrictions in the name of public safety and rights of consumers; increased competition in foreign markets; and the acceleration of national and international standardization. These trends, and undoubtedly others, have led to renewed demands for more and better information on the temperature-related properties and behaviors of metals."

Described as "the most significant organizational event in the decade" was the creation of the Metal Properties Council (MPC), which had been approved in principle ten years earlier by the Joint Committee and was later voted as the third sponsor of the Joint Committee. The primary objectives of the MPC during its initial decade were defined as the identification of needs for data on the engineering qualities of metals and alloys; the collection and evaluation of such data; and interactions to make that data readily available through reports and publications.

The most significant work of the Joint Committee during this era was the technical effort put forth by working panels and standing subcommittees. The Steam Power Panel, for example, represented a graphic cross section of trends and highlights in this field. During the first four decades of the Joint Committee's existence, steam power temperatures had risen from 725°F to 1,200°F, with a leveling off of most new plants in the range of 975°F to 1,025°F. "These advances in temperature created a flush of technical activity in the Steam Power Panel as problems were encountered and resolved." Termination in the upward trend of temperature in fossil-fueled plants, plus emphasis on nuclear-fueled power plants operating at much lower temperatures, brought about an abrupt change of emphasis in some areas of the Panel's activities.

The panel also sponsored and held workshops described as "providing one of the most useful functions of the modern Joint Committee." Among these were investigations into liquid coal-ash corrosion, in collaboration with the ASME Research Committee on Corrosion and Deposits; studies of boiler repair procedures; a review of papers of the ASME Research Committee on High Temperature Steam Generation; and investigations into fracture toughness as related to steam generation.

The Chemical and Petroleum Panel was active in many subject areas that directly affected the development of boiler and pressure vessel codes and standards. Characteristic were workshops on creep and creep-rupture strengths of austenitic stainless steels; cast corrosion-resistant high-strength steels; residual life in heater and reformer tubing; and operating experiences involving elevated-temperature alloys.

The work of this panel emphasized the special interest during this era in low-alloy structural steels, cast stainless steels, hydrogen effects, and coal-gasification. Another area under study, which indicated a trend at the time, related to the properties and changes that occurred in various metals and alloys when subjected to low and cryogenic temperatures. Much activity was focused on obtaining data for steels, titanium alloys, aluminum alloys, and composites, particularly in the matter of fracture toughness, fatigue, and suitability for the storing and transportation of liquefied natural gas (LNG).

The activities of the ASTM-ASME-MPC Joint Committee during this decade, and in the years that followed, were not only of inestimable value to the Code Committee in keeping pace and updating rules and specifications, but chronologically they served to record the trends that were significant in the very history of the Code itself.

CHAPTER 18

A CENTURY OF PROGRESS

ASME Enters its Second Century

In a lecture in 1987, entitled "Engineering and Safety: Partners in Progress," William F. Allen, Jr., Chairman of Stone & Webster, stated that "life can never be made risk free. That simply is not in the cards. Clearly we must continue to do our best to achieve the highest degree of safety consistent with a reasonable expenditure of human and material resources. But we must, at the same time, recognize that perfection in engineering — as in every other aspect of human endeavor — will always elude us. We should, nevertheless, continue to strive for it in a rational way."

Despite the fact that "providing for safety is the very essence of engineering," he explained, an attitude that is completely necessary to the establishment of codes such as those published by ASME, engineers still have to apply their experience and exercise their judgment in carrying out the design process. A certain percentage of failures are all but inevitable, otherwise we would never have had headline making accidents like the collapse of the Tacoma Narrows Bridge 50 years ago, the more recent failure of the bridge on Route I-95 in Connecticut, the collapse of the runway spanning the lobby of the Hyatt Regency Hotel in Kansas City in 1981, or the *Challenger* Space Shuttle disaster. Singly or in various combinations, inappropriate design, structural defects, and human error account for failures, major and minor alike, and usually get far greater attention in the press than technical achievements unaccompanied by dramatic moments.

Allen cited the Three Mile Island nuclear plant accident as what might have been a catastrophic disaster had not the safety system functioned and minimized the dangers of fire, radiation, or explosion, preventing casualties though not the severe economic losses. Although it demonstrated graphically that nuclear power, as exemplified in the water reactors of the Western world, is safer than had been thought and "far safer than most other activities in modern society," the distressing aftermath has been the increasing public fear of nuclear power. As Allen emphasized, "a minor radiation leak at a nuclear plant, for example, that is well below allowable limits, is usually a topic for national coverage on both radio and television." By contrast, even though nearly 200 lives were lost when a ferry capsized in Belgium after leaving port with the bow doors open, no one suggested that increased attention be paid to ferry safety in the United States.

Thus it has been for the past 75 years with the development and installation of the ASME Boiler and Pressure Vessel Code. The achievements, most notably in the prevention of casualties and property damage, largely go unnoticed by the public. Yet when an explosion occurs, or any other form of serious accident, the critics come knocking on the door to find a scapegoat and damn the Code for not having created rules and specifications that were foolproof.

This is perhaps all the more ironic in light of the historic fact that the founding fathers of the Code dedicated themselves and their heirs to establishing specifications and rules that would result in boilers that were constructed in "as nearly perfect a manner as possible."

In the middle of the 1980s, the nuclear power industry found itself in what was described in one editorial as a "good news/bad news" joke. The good news, according to the American Physical Society, was that the amount of radioactive materials that might be released in a severe reactor accident would probably be lower than had previously been estimated. The bad news, from the Nuclear Regulatory Commission, was that "there is almost a 50/50 chance of a core meltdown somewhere in America within the next two decades."

Taking the "bad news" first, using a typical core-melt probability of one in 3,300 years, based on risk assessments for two dozen sites and plants, the NRC calculated that there was a 45 percent chance of such an accident in a total of 100 operating reactors over a period of 20 years. However, it was pointed out that "only a small fraction of so-called core-melt accidents would have major off-site consequences."

Referring to accidents that were not meltdown related, the calculations for the amount of radioactivity that would be released were significantly lower than what had been predicted by the landmark Reactor Safety Study reported in 1975 because of three factors. First, reactor buildings had proven to be stronger than originally thought. Second, physical and chemical facilities that were designed to contain dangerous radioactivity within the reactor buildings had been greatly underestimated. Third, additional facilities such as suppression pools and auxiliary structures had been designed that would further neutralize the concentration and spread of radiation.

A significant aspect of the increasing reliability of safety factors at nuclear plants in the United States has, of course, been the development of the pertinent ASME rules in Sections III and XI of the Code, relating to the construction and inservice inspection of nuclear power plant components. ASME has taken the lead role not only in these current areas of operation, but in developing specifications and guidelines for the extension of nuclear plant life. In 1985, the ASME Board on Nuclear Codes and Standards (BNCS) discussed the need for coordination among all organizations engaged in writing codes and standards that might affect the life of nuclear facilities. It was pointed out that many *non*-nuclear codes were also vital since the effects of age-related deterioration in those areas could be just as vital as in areas specifically designated as "nuclear."

The ASME Board on Nuclear Codes and Standards had three basic functions: to administer ASME activities related to codes, standards, and accreditation programs applicable to nuclear facilities; to assess the need for further codes; and to recommend ASME policies and relationships with other groups in the nuclear field, including regulatory agencies. While it was not intended that the Society should try to develop any standards that traditionally belonged in other jurisdictions, it was mutually agreed by all participants that it was natural for the ASME to act as a prime coordinator. As perhaps the primary organization writing standards in the nuclear field, the ASME was thus recognized for its accomplishments during the history of the Boiler and Pressure Vessel Code, most notably in Sections III and XI, yet not excluding other sections that were applicable to nuclear development, such as Material Specifications, Pressure Vessels, Welding, and Nondestructive Examination. It was widely recognized that substantial contributions were made by B31 in the nuclear piping area and by Performance Test Codes in the matter of pumps and valves.

In taking this leadership role, the Society enjoyed the support and endorsement of the American National Standards Institute (ANSI), specifically its Nuclear Standards Board, a standing committee that did not write standards but had planning and coordinating responsibilities in the nuclear field. Thus it was that the Boiler and Pressure Vessel Code was recognized in a unique way for its historic role in the evolution of codes and standards.

Critics of codes and standards in the nuclear industry tended to base their objections on two claims: *first* that the response to the needs were not "timely" enough, often taking what seemed like an interminable period to move from the recommendation stage to final approval, and, *second,* that the rules were more restrictive than necessary. Supporters of existing code procedures countered these critics by pointing out that, while there might have been some vestiges of truth to complaints like these, vital and enduring benefits such as the following were often overlooked.

- Any standards-writing organization that expected accreditation had to meet the requirement that its code committee be composed of *balanced* representation from the industry, as well as provide for public review and comment.
- No matter how requests for code changes originated, they had to be reviewed first by the committee at the lowest tier, on the "firing line," to determine if changes were indeed warranted and supported by reliable data.
- In all, technical revisions had to undergo at least five levels of review before being approved. This procedure, while admittedly slow, assured that all needs concerning the safe operation of nuclear

facilities and parts were given the most careful and thorough consideration.

- Proposed changes could be denied for cause at any point along the way. However, they could also be added to or improved upon by a subgroup with exceptional expertise or vision in the subject area under consideration, thus providing a positive input that could be valuable.

A major contribution of the Code Committee in the late 1980s has been directed at programs for the extension of plant life at nuclear facilities, while still maintaining all of the basic factors of safety and reliability. Structurally, this trend has been apparent in such matters as the organization of a Special Working Group on Plant Life Extension, providing data and recommendations. The greatest effect has been upon the makeup of Section XI of the Boiler and Pressure Vessel Code, and will continue to be.

The nature and degree of change pose sensitive problems because they must not only take future nuclear developments into account, but must avoid penalizing existing utilities that do not intend to apply for license extensions for their nuclear plants when existing authorization comes to an end under the traditional 40-year limit.

This is the substance of one of the major areas of Code activity of the middle and late 1980s, a trend that will not fade or diminish, despite the questionable position of nuclear energy in North America at the end of the 20th century. Yet the Committee and all those concerned with nuclear codes and standards will constantly have to deal with the multitude of problems behind the headlines, of which the following is a characteristic example: DEFECTIVE BOLTS FOUND IN HALF OF NUCLEAR REACTORS.

This headline in *The New York Times* on June 19, 1988, pointed to the kinds of problems faced by ASME and other organizations concerned with codes and standards in the nuclear field. A "top official" of the Nuclear Regulatory Commission had informed a Congressional subcommittee that more than half of the nation's 109 nuclear reactors had some substandard studs and bolts in safety-related locations.

However, he went on to testify, none of these suspect bolts had so far posed a safety threat. The problem stemmed from the failure to test crucial steel bolts that suppliers had sold to users as being resistant to bending under stress and had been falsely labelled "high tensile."

It was reported that the subcommittee had for the previous three years pushed for a crackdown on foreign imports that were too cheaply made. The subcommittee, furthermore, criticized the Nuclear Regulatory Commission for not requiring utilities to make stricter inspections when they received components — especially those of foreign make — for use in reactors. "The simple fact," charged one Congressman, "is that the NRC is dragging its feet. Bad bolts can be like a virus that can weaken our infrastructure and can cause our collapse."

One outcome of this revelation was that the NRC would look into the matter and would soon be "exercising a rather heavy hand" in pressing civil claims and criminal charges against "known counterfeiters of bolts, pipes, and other steel parts used in nuclear reactors."

Was this incident perhaps a cogent example of the weaknesses of bureaucratic intervention in the regulation and administration of codes and standards? That was a question, among many, that would have to be weighed carefully in the efforts to preserve the past, protect the present, and plan the future.

One of the most active fields of development as the ASME Code approached its 75th anniversary was certainly in the field of nondestructive examination, the subject area covered specifically by Section V. These activities tied in with, but were by no means limited to, the program to extend the life of nuclear plants and components.

Although many applications of new nondestructive testing equipment and techniques were applied to materials and equipment that were outside the specific realm of boiler and pressure vessel codes, sooner or later they found more direct uses of interest to Code reference. One paramount advance of the 1980s was acoustic emission, which paradoxically was a natural phenomenon that had been observed for many centuries. The first commercial use of acoustic emission, referred to in the industry as AE, may have been by potters who relied on the audible cracking sounds of clay vessels cooling in the kiln to detect which ones had flaws and would eventually crack. More closely related to today's applications of the emission was "tin cry," a sound audible to the naked ear that was made by tin that had been smelted. In the 16th century, it was noted that iron made certain sounds during the primitive forging processes of the day that could be recognized as signals of what would later be seen to be structural defects.

These were really prototypes of the type of acoustic emission that began to emerge in the 1980s as a sophisticated and reliable method of nondestructive examination that would have a diversity of industrial applications. The use of AE was to become increasingly important in the field of power, for testing components used by public utilities and critical elements in nuclear power plants that were difficult to reach effectively by other methods of examination. The acoustic emissions of metals under widely varying conditions and in the broadest kinds of applications was so startling that some of those who experienced the phenomena have referred to a future in which "intelligent metals" will one day alert users to potential zones of failure through this bizarre means of inanimate communication.

One example of the use of AE in the late 1980s came about because of the continuous cases of failure of hot reheat steam lines because of creep damage. The problem stemmed from the unique combination of high temperature and stress in this type of piping, aggravated by thermal fluctuations that were characteristic of this kind of an operating cycle. It had long been known that creep was an ongoing process that tended to progress at an ever-increasing rate as metals approached the point of failure. The problem was that conventional methods of testing areas that were suspect, while revealing the conditions of metals at the times of the inspections, did not indicate the *rate* of deterioration or provide forewarnings of incipient failures.

Acoustic emission evolved as the most reliable method of continuous, in-service monitoring of steam lines, whose normal noise levels were low enough so that they did not cloak the acoustical messages of the testing instruments. According to researchers in this field, AE held great promise for many other fields of testing as well, despite the fact that it was only just beginning to come into its own.

"Acoustic emission technology, when properly applied," reported the American Society for Nondestructive Testing in the winter of 1988, "enables the nuclear power industry to know the mechanical condition of critical equipment bearings and to evaluate the leak conditions of system valves, both gaseous and liquid."

Hand in hand with new achievements and expectations in nondestructive testing went rapid progress in computer design and applications. In 1980, a series of engineering papers had been researched and published under the title, *Advances in Computer Technology.* One section of the many pages of contents was particularly revealing and related to computers in energy systems. Although computers and all kinds of automated systems had been in operation for many years prior to that time, 1980 marked the advent of a decade in which computers would become all but indispensable in those fields covered directly by the ASME Boiler and Pressure Vessel Code.

Characteristic papers were those describing the use of computers for such purposes as measuring the velocity and pressure field in once-through steam generators; simulation of a total energy system; studying the aerodynamics of steam furnaces; modeling hot water systems; analyzing solar heating and cooling systems; and minimizing the operating costs of power boilers.

The last-mentioned subject was based on the concept that "since the power boiler is the key energy user in conventional fossil-fuel steam generating plants, any reduction in fuel consumption via computer control can mean substantial savings." The study by the authors considered the practical application of a computer to operate a set of power boilers supplying a common load, the objective being to maintain control while minimizing the costs of operation. Since boilers in the United States were then consuming about 20 percent of the nation's energy usage, conservation through computerization could help to reduce industry's dependence upon scarce or costly fuels. It was estimated that this approach might effect fuel savings of from three to five percent.

Tests like this one and other research in depth proved during the 1980s that the computer and its software could be adapted for use in almost any field relating to the Code and its applications. One program, for example, found a practical application in calculating stress in piping systems. Another was programmed for the design, analysis, and rating of both horizontal and vertical pressure vessels in accordance with the rules and specifications for Section VIII of the Code. It could also calculate head and shelf thickness and optimize nozzle sizes, among other spheres of activity. And yet another program had been perfected for the storage and retrieval of welder qualification records. It could accommodate a number of codes and standards, including Section IX of the Boiler and Pressure Vessel Code. It made possible single or multiple searches to obtain data about relevant codes, inspecting authorities, specifications, materials, and almost any other information that might be needed, instantly and reliably.

A major computer phenomenon of the decade became known by the rather unfortunate acronym, CAD, which translated into Computer Aided .Design. This was touted broadly as the problem solver par excellence in the design of complex industrial systems, and particularly useful for power plants and process plants characterized by flows of fluids and gases in pipes connected to boilers, pressure vessels, and units of machinery. About 50 percent of the construction cost and more than half of the design cost of a power plant was traditionally budgeted for designing the piping network and arranging the locations of equipment and components for the most efficient operation. These were tasks for which computers and software had proven themselves in the past, very effective when it came to charting flow diagrams.

One important breakthrough in 1989 was the introduction of the Boiler and Pressure Vessel Code on CD-ROM, (Compact Disc-Read Only Memory). The entire Code, including all texts, tables, and figures, was for the first time made available on a single disc that is less than five inches in diameter. Despite its feather weight and tiny size, the CD-ROM disc has such enormous capacity that it can store approximately 275,000 pages of the size you are now reading. All the resources needed for information, data checking, and decision-making are at the user's finger tips, such as Code sections, interpretations, cases, referenced ASME standards, and even charts and graphs. In the case of visuals, they can be instantly enlarged and zoomed to bring out the minutest details.

Sophisticated systems of data processing, along with advances in testing procedures, have wrought great changes in many of the old, traditional spheres of activity that had been progressing all the while but at rates that were sometimes painfully slow and ponderous. Since the early 1930s, for example, the ASME Research Committee on the Properties of Steam had been providing reference data on thermal properties of water and steam for the American power industry. Much more recently, it began coordinating its research programs with the International Association for the Properties of Steam (IAPS). Together, the two groups have escalated their release of continuing formulations, equations, and other numerical data for chosen properties over wide ranges of temperatures and pressures, along with applicable tables. These properties of steam and hot water have been used successfully in the design of boilers and turbines and in accurately calculating such desired functions as efficiency and control.

Another milestone in recent Code history has been the evolution of techniques and equipment for monitoring the performance of older, fossil-fueled power plants, recognizing the need to upgrade them in light of the cancellations and postponements that have been plaguing the nuclear field. Although performance monitoring goes far back in the history of power plant operation, it was little more than a routine procedure until recent times when consumers began to object strongly to having utility companies pass along to them cost increases which they felt were unnecessary.

One breakthrough in the monitoring and testing of power generation equipment during the past decade relates to the undesirable presence of vibration. According to a technical report from the Babcock & Wilcox Research Center in Alliance, Ohio, in 1985, "Vibration measurement and analysis are of critical importance in the development, installation, and maintenance of fossil and nuclear power generating equipment. The multiplicity of products involved (reactor vessels and internals, steam generators, boilers, coal pulverizers, and fans, among others) combined with the harshness of the environments in which they must operate, demand that sophisticated equipment be used to make the measurements and that highly trained personnel interpret the results."

One reason why the 1980s was to become a period of active research in the power industry lay in the close relationship between technological evaluation and good management. This was an era when public utilities were becoming

216

far more competitive than they had ever been before and when failure to maintain a healthy economic condition could lead to poor stockholder relations and sometimes public outcry. Research did not solve all the problems, but it certainly paid its way in the more efficient generation and supply of power.

CHAPTER 19

CONCLUSION
Education, Planning, the Future

E ngineering's role in the future can go only one way, toward a central role in the planning and management of society. Along this direction lies an opportunity for the revival of engineering greatness. . . ." So states the introduction to a study, "Forces Shaping the Future of the Engineer and the Engineering Profession," prepared for the Board on Issues Management of ASME by J. F. Coates, Inc., Washington D.C., and published in June 1987.

The study cited artificial intelligence, computer and telecommunication links, robotics, materials, biotechnologies, geotechnologies, and energy as "key areas of development," several of which directly affect the future of the Code.

Two sections of the study were of primary importance in this respect. The first covered the professional aspects under the thesis "keeping up with the pace of technological innovation is engineering's toughest challenge." Like Alice in Wonderland, "engineers will have to run faster just to stay in place, let alone keep up with the accelerating pace of new technological applications."

Common to almost all developing technologies, and most certainly to subject areas covered by the Code are the basic trends toward:

- *miniaturization* — with components that are becoming smaller, cheaper, more powerful, and more efficient;
- *complexity* — with more technology crowded into comparable units, large, small, and miniscule alike;
- *modularity* — permitting the replacement of units as separate "pods" when they become worn out or need to be updated;
- *integration* — of product design, production, and marketing by planning systems that produce all three;
- *safety/quality/durability*— with the application of technological advances that can assure all three factors.

The second section of import related to the matter of regulation and its impact on technology and engineering. "Regulation is likely to flourish," reported the Coates study, "as structural change in the economy, international competition, and new technology create complexities and instabilities." Among the specific areas where regulation was seen to have emerged or be emerging, and which are covered by ASME codes and standards, were energy use and production, conservation, and product safety.

As a result, "engineering societies may be more involved in providing information and analysis to regulators as they move into the tentative and experimental management of new and more complex products and processes."

221

The study made the forecast that more and more engineers and divisions of engineering societies would be called upon to design effective regulatory experiments; that engineering societies would become more anticipatory and more sophisticated in their understanding of technology/ regulation relationships; and that engineers would work more effectively with legislators and other decision makers, especially at the state level, as experts and consultants in the regulatory process.

The Codes and Standards Department of ASME has a policy of continuously assessing new technological fields or existing ones that show signs of radical evolution, always well aware that technological advances provide new materials and concepts for the development of codes and standards and the administration of accreditation activities. Committees try to anticipate the nature and degree of changes that will take place, since some of them — for example, the introduction of new materials — can result in delays in the publication of related codes. Typical of new projects recently cited are:

— computer-aided design standards, being worked out with the National Bureau of Standards;

— assessment of pressure vessel research to determine the extent to which existing requirements in Sections III and XI should be simplified, clarified, improved, updated, or eliminated;

— reinforced thermosetting plastic pressure vessels;

— high-pressure systems for pressure vessels;

— piping systems for pressure vessels for human occupancy;

— welded aluminum storage tanks;

— issues of nuclear power plant aging and life extension, a project in which ASME has taken the lead.

"The nuclear industry has changed markedly in the past ten years," reported the ANSI Nuclear Standards Board (NSB) in a white paper on policy in the spring of 1987. "These changes suggest that those involved in the development of voluntary consensus standards take time to look at where we are, where we have come from, and where we are going. From this perspective, the nuclear standards program needs to be critically evaluated to determine what it should be doing today. Is it serving its users? Is it using resources wisely? Does it satisfy a continuing need?"

Stating that the structure, groundwork, and basic strategy of the standards effort were laid in the 1960s and 1970s, the Board emphasized that this was a period of strong growth in which standards were critically needed. However, it added, "things have changed and the situation is different today. . . . the

222

industry is principally looking at operating plants — not design and construction. The watershed of the 1970s is finished."

NSB forecast that the nuclear industry would continue to change drastically in the 1990s and beyond. In line with this, "the nuclear standards program must be focused and managed in such a way as to be on the cutting edge of the nuclear business." To accomplish this future objective, three subjects had to be addressed. *Existing standards,* to ensure that they are still technically accurate, consistent, and needed; *new standards,* to limit the development of any that might be questionable and to ensure that all are the proper solutions to indentified needs; and *maintenance of standards,* to make certain that they continue to be technically accurate, practical, and workable.

Among the Board's goals for the future were to promote the maintenance of a single, consistent set of American National Standards for the nuclear industry so that there would be no conflict between standards as to requirements and no unnecessary duplication of effort; to maximize the close liaison and interface between the nuclear standards-writing community and the users of standards; to assess all new nuclear standards projects by a broad spectrum of the nuclear community; to assess needs and priorities; to evaluate all nuclear standards projects not completed within two years after their initiation; and to be aware of and watch the development of issues regarding the life extension of nuclear plants.

Although the Board set out to aim at a single consistent *set* of standards, that is not to infer that anyone has the mistaken idea that there is a simple across-the-board formula. John F. Harvey interpreted the difference when he spoke about one specific type of pressure vessel. "Will the ASME Code ever reach the point where it will have a single uniform approach or formula or rule for the construction requirement of a cylindrical vessel subjected to internal pressure? Hardly, and for the very simple reason that it must continue to fulfill its mandate to provide rules for the safe construction of all kinds of vessels, types of manufacturers, and kinds of materials."

There is no single all-inclusive panacea. The Code must make available suitable methods and means for the use and implementation by the small fabricators of low-pressure heating boilers, vessels, or tanks who do not have professional engineering staffs at their disposal. This segment makes up by far the major number of vessels built each year. Yet the very same ASME Code body must also provide for the safe construction of the most advanced and sophisticated vessels employing exotic materials in hostile environments, subjected to ultra-high pressures, and constructed by large, well-staffed companies backed up by extensive computer facilities and research organizations.

223

Although it has been said many times in many ways that the key to tomorrow lies with the research of today, the public is confused by the meaning of the word "research" and has an unclear image of its function. "Relatively few people, other than those of us who do it or who use its results really understand what engineering research is," said Daniel C. Drucker, Graduate Research Professor in Engineering Sciences at the University of Florida and a past president of ASME, in an address in December 1986.

"While basic scientific research studies the world of nature, fundamental engineering research attempts to expand our understanding of the technological world. The dominant motivation is to achieve knowledge and understanding that will lead to practical advances in the field. A long sequence of further steps in applied engineering research, development, design, and production is normally required before an engineering advance is ready for public use."

One area of R&D that has been repeatedly cited as part of the "wave of the future" has been NDE — nondestructive evaluation. The key to the continued preeminence of the United States in aeronautics, says a technical study from the Rockwell International Science Center, "is to be found in the research, development, and application of a group of revolutionary technologies in the areas of propulsion, numerical and symbolic computation, laminar-flow modeling, and advanced materials and structures." Materials science and engineering, and especially the discipline of nondestructive evaluation, will play a major role in this industry. The need to ensure the reliability of materials has always been critical and will be even more so as advances are made in equipment and materials, many of them exotic by comparison with traditional materials of the past.

As conventional metals, alloys, ceramics, and polymeric composites approach the limits of their capacity for energy and aeronautical applications, new materials must be perfected that will withstand the challenges imposed for fatigue and thermal strength. The diversity of materials and structural applications will dictate the use of the most advanced nondestructive techniques for their examination and testing. Computer-based ultrasonic systems, for example, will be used to inspect components, functioning far more quickly than their non-computerized predecessors. Real-time radiography is another technique for evaluating composite materials, along with fiber-optic borescopes, dye-penetrants, X-ray, and magnetic devices — all of which are in use today but will be greatly enhanced in the future.

"Once reserved for detecting microscopic flaws in structures after they had been built, NDE techniques are now being applied to materials at all stages of production and use," reports the Center for Nondestructive Evaluation at Johns Hopkins University. Recognizing the needs, the ASME itself formed its own engineering subdivision to assess the field of nondestructive evaluation as a vital function in the evolution of codes and specifications. NDE has matured into a multidisciplinary operation, whose techniques and methods require a knowledge of physics, acoustics, electromagnetic theory, optics, radiography, and other disciplines that will be applicable in the future.

A Sampling of Characteristic Areas in Which Code-Related Developments Will Affect the Engineering Future

Although it is difficult to evaluate the speed or scope of progress in technological fields because of the many complex factors that are influential, the following examples represent a cross-section of what to expect, particularly in those subject fields that members of committees on codes and standards must be knowledgable about as they tackle the many challenges on the way to writing and publishing new segments of the eleven sections of the Code.

The Factory of the Future

Computer-integrated manufacturing control systems of the future will provide engineers in the 1990s with "a palette of tools" for planning, scheduling, and controlling operations. Once a manufacturing objective has been defined, such as the fabrication of steam piping, systems engineers will be able to select from those tools the modules they need to build a control network for particular products or product lines. As the authors of a report on *The Factory of the 90's* envisioned it, a truly flexible system would be controlled by a distributed computer grid that could be adapted to specific objectives. It would consist of four tiers, including administrative guidance, for which business computers would suffice; supervisory control that required high-performance business computers; machine monitoring, using microcomputers; and direct machine control, supervised by dedicated programmable controllers.

This computer arrangement would be so designed that, upon any kind of failure at higher levels, the operations at lower levels would continue uninterrupted. However, all levels would ideally switch automatically to back-up networks upon receiving signals that there was any interruption of the primary systems.

Freight Train of Tomorrow

Imagine a train in nearly constant motion, with motive and load-carrying units designed as an integral system rather than as separable cars, plying continually between selected locations, loading and unloading cargo swiftly by means of mechanized bulk-handling techniques.

The adaptability of the railroads should not be underrated. They are now on the verge of introducing what may be the technological white knight that will return the industry to the dominant position in transportation it once enjoyed. This new concept is called the *integral train.* Simple in concept, it will consist of a single, integral unit with inseparable cars and logically spaced power units. The motive units would be lightweight, high-speed dieselengines, podded like aircraft engines for rapid replacement. A long train might be powered by two or three locomotive units. Since the cars and engines would be integrated rather than using conventional couplings, the design would eliminate the slack that lowers the efficiency of conventional trains while changing speeds. Such innovative concepts would necessitate changes in codes and specifications that would be radical by comparison with ones evolved for other forms of transportation.

Solar Energy

Although solar energy is basically very simple and can be readily harnessed by amateurs for limited supplies of heat in and around the home, its commercial applications have been slow in developing. Today, only the tiniest fraction of the available solar energy has been harnessed, largely because of the problems of collecting, converting, and storing it efficiently. The problem has been that, although sunshine is free, the means of collecting and using its energy have been more expensive than for almost every other form of energy available, including oil, gas, coal, and other conventional fuels. Solar energy has been described as "a field for visionary inventors and entrepreneurs." Yet it will certainly take its place among widespread energy systems by the end of the 20th century.

Power in the Sky

For strategic defense, we may one day see power conversion units situated in space, which might require their high-power use for only an hour at a time but which would have to remain battle-ready for between 10 and 20 years. Nuclear and chemical combustion methods are being investigated as sources of power for the space platforms. Both would power gas turbine engines, which offer the best means for exploiting the high-temperature potential of both nuclear and chemical combustion. The use of mature gas-turbine technology and existing materials would result in highly reliable power conversion units

(PCUs) capable of meeting the tough requirements. Because the efficiency of the proposed system would be improved through operations at high pressures and elevated temperatures, the availability of materials becomes a dominant issue. The open-cycle chemical turbo generator promises to be the simplest and lightest system that could be available for providing power. The power conversion systems needed will not be deployed for another 10 or 15 years.

Heat Engines With Built-In Reliability

Two promising technologies for the direct conversion of energy are thermionics and thermoelectrics. The small size, low weight, and lack of moving parts in these power units make them suitable to remote or harsh environments.

Thermoelectrics is by now a proven technology, used extensively in the space program, but thermionics, whose practical development only began in the mid-1950s, had a lapse in R&D for ten years and the program has only recently resumed. Potential applications for the two technologies include instrumentation, exhaust-gas energy recovery, refrigeration, and topping power systems. Both are particularly applicable to hostile or distant locations where cost is less important than reliability. However, as the cost of these units decreases and their efficiency increases, they will find more commercial applications on earth as well as in space.

A thermionic energy converter is perhaps the most direct form of heat engine that exists. TECs have no moving parts, require no maintenance, and are inherently rugged. Even within the core of a nuclear reactor, a TEC unit can last from five to ten years. Solid-state semiconductor devices, by means of the principle of thermoelectrics, can directly convert heat energy into electricity without the need for dynamic machinery such as a turbine or generator. Operating as thermocouples, such devices are used extensively in the conversion of low heat energy sources into electrical signals proportional to the temperature.

"The current level of interest in both thermionics and thermoelectrics is very high," says the Georgia Institute of Technology, which has undertaken a considerable body of research in this field. "While resources shrink and the worldwide demand for energy grows — along with the cost of conventional fuel supplies — the need to find alternative sources of power becomes more pressing. Meanwhile, improved conversion efficiencies and lower costs promise to increase the number of terrestrial commercial applications for both technologies."

The Coming Geothermal Buildup

The Geothermal Energy Symposium, sponsored in January 1988, by the Geothermal Resources Council and the Petroleum and Advanced Energy Systems Division of ASME, provided an opportunity for engineers to get an overview of the state of the art and an image of potential future developments. This was the first time that geothermal energy had been included among the symposiums at an Energy-Sources Technology Conference and Exhibition (ETCE), sponsored by ASME and seven other engineering societies from the United States and Canada. Among the subjects covered were advanced concepts, research and development programs, and the state of the Magma Energy Project to assess the engineering feasibility of extracting thermal energy directly from crustal magma bodies, such as Kilauea IKi lava lake in Hawaii.

"The estimated size of the U.S. resource suggests a considerable potential impact on future power generation," reported one Symposium paper. "Magma intruded into the crust is the heat source for geothermal reservoirs. Magma also represents an energy resource that is both much larger and of higher quality than the geothermal resource. It is estimated that this resource, within the upper 10 kilometers of the crust, in the United States ranges from 50,000 to 500,000 quads — larger than the current estimate for the nation's fossil reserves."

Geothermal energy is recognized today as the only one of the so-called alternate renewable energy resources that has proven itself technically and economically and that is already commercially available for electric power production in 17 countries. Estimating that the annual growth in geothermal production will be about five percent, the annual output of slightly more than 5,000 megawatts would rise to almost 6,400 megawatts by 1992. The United States will sustain its present leadership in this energy field, it was estimated at the Symposium, with about 140 units on line in the early 1990s, accumulatively producing some 2,650 megawatts annually.

Among them, the four leaders — the United States, Italy, the Philippines and Mexico — will produce about 85 percent of all the geothermal power capacity during the 1990s.

The Persistence of Wind Turbines

The future of wind energy as a commercial source of electricity has been said to be "dependent upon a cascade of technical and economic issues, as well as upon fossil fuel rising." Out of this mixed bag of dependencies, the technical aspects are the ones that are by far the most clear. Wind energy, in terms of the fundamental resource, is sufficiently abundant. Wind turbines to extract energy from the wind are technically feasible and there already exists an industry that designs, builds, and markets various sizes and models of the end product. Wind-turbine development went through two distinct generations of machines during the decade from the mid-1970s to the mid-1980s. The third generation, anticipated for the beginning of the 1990s, will be cheaper, yet more durable. "If third-generation wind turbines prove to be as cost-effective in practice as they are on paper," said one report, "wind energy will take its place as a viable commercial enterprise."

Ocean Engineering: Wave and Thermal Energy

On an average day, it is said, the equatorial oceans receive an amount of solar radiation that is thermally the equivalent of hundreds of billions of barrels of oil. But it has also been emphasized that "the engineering problems in converting this energy seem correspondingly vast."

Electrical power can be harnessed from two major forms of ocean energy: ocean thermal energy conversion (OTEC), which utilizes the difference in temperature between the warm waters on the surface in southern climes and the cooler waters deep below the surface, and energy from waves and immense movements of water, whose kinetic and potential energy can sometimes be extracted.

The greatest engineering challenge in OTEC is that equipment must be of gargantuan dimensions, handling flows of millions of gallons of water per minute, equivalent to rates as huge as those of major rivers. But the energy storage capacity of seawater is sufficient to keep OTEC plants in continuous operation around the clock. The benefits derive from the fact that if plants could convert only a fraction of one percent of the region's available heat to energy, they would provide more than 20 times the installed electrical capacity in the United States, more than 600,000 megawatts.

As for wave energy, there have been designed more than a dozen generic types of wave/energy systems. Some extract energy from surface waves, others from pressures fluctuating below the surface. Some systems are fixed in position and function as waves pass through and by them, while others move with the motions and currents. Wave energy systems are designed for peak-power situations, in contrast to OTEC facilities that supply base-load power. Although the minimum output requirement for commercially viable wave energy plants is smaller than for OTEC plants and could be cost-effective when producing as little as 0.5 megawatts of power, construction of commercial facilities seems to be some years in the future.

Coal — an Old Technology With Yet Another New Future

With all the focus on advancing technology and radical breakthroughs in the energy field, it seems on the surface to be something of an anachronism to bring up the subject of one of the world's oldest commercial resources: coal. Yet that is exactly what R&D in the late 1980s has been doing, in effect seeking two objectives. The first, to which most Federal funding has been proposed, in increments of some $500 million a year, is "clean coal," to support the U.S./Canadian venture to fight acid rain. The second, a continuing effort by the steam boiler industry, has been the improvement of various coal-related fuels and firing facilities to make coal more cost-effective as well as environmentally compatible.

As one coal-research engineer expressed it, "High-tech devices, exotic new fuels, and revolutionary methods for supplying cheaper, more reliable power come and go, but coal ambles on forever. If you want a surefire prediction, I can tell you right now that, come the year 2000 A.D., *coal* will be at the forefront of 21st Century R&D."

INDEX

Association of American Railroads, 107
Association of American Steel
    Manufacturers, 38
Assyria, 67
ASTM/ASME Joint Committee, 84
ASTM/ASME Joint Committee, 50th
    Anniversary, 1975, 204
Atlantic Refining Company, 104
Atomic Energy Commission (AEC), 167
Atomic energy. *See* Nuclear energy
Atomic Industrial Forum, 186
Atomic Safety and Licensing Board, 173
Atoms for Peace Program, 3
Automatic coal stokers, 143

Babcock & Wilcox Company, 27
Babylonia, 67
Baldwin articulated steam locomotive,
    204
Baldwin Locomotive Works, 18
Baltimore & Ohio Railroad, 18
Baltimore, Maryland, 125, 126
Battelle Memorial Institute, 86, 183
Battista, Giovanni, 13
Bent-tube boilers, 15
Berkness, Russell, 24
Best Friend of Charleston, The
    (locomotive), 17
Binary steam cycles, 89
Board of Inspectors of Steam Vessels,
    48
Board on Issues Management, 221
Boehm, William H., 32
Boiler failures and explosions, 21-28
Boiler Insurance and Steam Power
    Company, 22
Boiler Tube Manufacturers of America,
    39
Boiling water reactor, 166
Bolted flange connections, 98
Boston Edison Company, 3

Boston, Massachusetts, 27
Bowles, Francis T., 94
Brazing, 103
Brister, Paul M., 136
British Standards Institution, 78
Brittania (steamship), 16
Brittle fracture. *See* Embrittlement
Brown Paper Company, 93
Brush Electric Light Company, 15
Buel, Richard H., 48
Buffalo, New York, 149
Bureau of Explosives, 107
Bureau of Mines, 143
Bureau of Standards of the United
    States, 87

Calder Hall (England), 163
Canada: 78,79
Canvass Method, 7, 188
Carpenter, Rolla C., 32
Cartwright, Edmund, 67
Caspian Sea, 17
Caustic embrittlement. *See*
    Embrittlement
CD-ROM (Compact Disc-Read Only
    Memory), 215
Centennial History of ASME, 24
Center for Nondestructive Evaluation,
    224
Certificate of Authorization, 169, 202
Chaldea, 67
Challenger Space Shuttle, 209
Charleston, South Carolina, 204
Chattanooga, Tennessee, 204
Chemical processing. *See* Petroleum
    processing
Chicago, Illinois, 50, 56, 93-98, 110
Chuse, Robert, 77, 111
Cloud, R. L, 173
Coal gasification, 138, 143

234

Hydrolevel Corporation. *See* Hydrolevel Corporation Case
Hydrolevel Corporation lawsuit, 7, 209-17

"Improved Application of Coal Burning Equipment, The," 143
Industrial developments, nuclear. *See* Nuclear energy
Inquiries about Code requirements, 195
Inspection, 71, 78, 88, 97, 112, 121, 123, 125, 129, 151, 163, 166, 169, 194
Inspectors and inspection methods. *See* Inspection
Institute of Electrical and Electronic Engineers, 187
Institute of Metals, 99
Institute of Mining and Metallurgical Engineers, 99
Instrumentation, controls, and monitoring, 84, 154-55, 230-31
Insurance, 23, 78, 144, 151, 156, 169-170
Interborough Rapid Transit Company, 76
Internal pressures. *See* Pressure increase
International Association for the Properties of Steam (IAPS), 216
International Electrotechnical Commission, 7
International Nickel Company, 50, 175
International Organization for Standardization (IOS), 7, 62
International Symposium, 87-88
International Telephone & Telegraph Company, 201
Interstate Commerce Commission, 54, 105

"Investigation of Graphitization of Piping," 182-183
Iron Age, 133, 134-135
Iron and Steel Board, 24-25
Iron and Steel Electrical Engineers, 61
Iron World, 32-33
Ithaca, New York, 33

Jacobus, David S., 27, 83, 189
Jet aircraft, 88, 149
Jet propulsion and rockets, 165-166
Johns Hopkins University, 124-125, 224-225
Joint API-ASME Pressure Vessel Committee, 69-70, 94-97, 115-16
Journal of Commerce, 50, 51

Kaiser, Josef, 126
Kansas City, 116, 209
Kellogg, M. W. Company, 164
Kennedy Administration, 194
Kent, William, 24-25, 32-33
KilaueaIkilava lake (Hawaii), 228
Kjellberg, Oscar, 95

Langer, Bernard F., 112-113
Lardello, Italy, 4
Lehigh University, 185
Lenoir, Joseph Etienne, 18
Lester, Horace, 122,125
Gas (LPG), 105, 117
Locomotion No. 1 (locomotive), 17
Locomotive boilers, 88, 164
Locomotive codes. *See* Locomotives, steam
Locomotives, diesel, 226
Locomotives, steam, 17-18, 53, 154-155
London, England, 87-88
Los Angeles, California, 84
Low, Fred Rollins, 50, 80

Obert, C. W., 32-33, 37, 63, 80
Ocean Thermal Energy Conversion
    (OTEC), 229-30
Office of Management and Budget of
    the United States (OMB), 197
Ohio Power Company, 163
Ohio rules and regulations, 31, 44
Oregon State College, 141
Outdoor power plants *See* Power
    generation

Pacific Area Standards Congress, 7
Pacific Gas and Electric Company, 4
Packaged boilers, 153-54
Palmer, Lewis R., 61
"Parker Case," 193-94
Parker, John Clinton, 43, 193
Parker, Walter B., 46-47, 57-58
Parsons, Kansas, 76
Patriot's Point Naval and Maritime
    Museum, 204
Pearl Harbor (Hawaii), 149-50
Pearl Street Central Station (New York
    City), 16
Penetrant (dye) inspection, 129
Pennsylvania rules and regulations, 40
Petroleum and chemical processing, 85,
    104-106, 165-166, 175, 205
Philadelphia Centennial Exhibition of
    1876, 15
Philadelphia, Pennsylvania, 15, 43, 193
Pittsburgh, Pennsylvania, 32
Pittsburgh Testing Laboratory, 32-33
Pleasanton, California, 166
Polytechnic Club, 22, 22
Post-weld heat treatment. *See* Heat
    treatment *Power,* 57
Power boilers, 79, 214
Power conversion units, 226-227
Power generation, 51,153, 218
Power testing, 50,

Pressure increase, 83, 99, 103, 105, 107,
    117-18, 120-22, 133, 151, 178
Pressure Vessel Research Committee,
    97, 137-38
Pressure Vessels and Piping: Design and
    Analysis, 182
Pressure vessels, 69, 97-99, 103-116,
    137,138,139, 153, 165,167,
    173-174, 182, 215, 222
Pressure, external, internal. *See* Pressure
    increase
Pressurized water reactor, 171
Program evaluation, 108, 184
Prohibition Amendment, 51
Providence, Rhode Island, 15
Public utilities, 89, 193
Pulverized coal, 52, 75-76, 143, 162

Qualification of inspectors. *See*
    Inspection and inspection
    methods Quality Assurance,
    181-82

Radiographic testing, 72, 124-125, 129
Reactor Safety Study, 210
Reciprocating steam engines, 17, 56
Reed, Edward M., 22
Relief valves. *See* Safety valves, 27-28
Remote controls. *See* Instrumentation,
    controls, and monitoring
Rensselaer Polytechnic Institute, 45
"Report of the Boiler Code Committee
    of ASME," 37
Reporter(ANSI), 7
Research and development, 235-44,228
Riedel Publishing Company, 59
Riveting, 71-72, 93
Robinson, Ernest L, 86, 149
Rocket (locomotive), 17
Rockwell International Science Center,
    224

www.ingramcontent.com/pod-product-compliance
Lightning Source LLC
Chambersburg PA
CBHW050456190326
41458CB00005B/1315